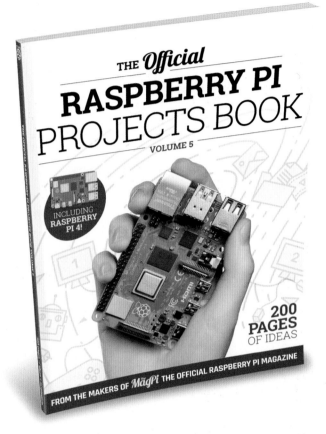

HELLO!

Can you believe that we're onto the fifth *Official Raspberry Pi Projects Book* now? Compiling a book of past projects, guides, and tutorials every year can be hard work, even if it is a lot of fun. For this volume, though, we once again had so many great projects to choose from that the hard work was trying to squeeze in as many as we could over these 200 pages!

I'd love to say that doing a Raspberry Pi 4 edition of the *Projects Book* and filling it with such outstanding projects from the wider community was to celebrate reaching the big number five; however, it's more of an indication of just how much great stuff the maker community has been making over the last year or so.

If this is your first time using a Raspberry Pi, you'll find some very helpful guides to get you started with your Raspberry Pi journey. Once you've done that, I'd like to extend an invitation to check out some of the fun projects and tutorials in the rest of the book, and try to make something truly amazing for yourself.

Who knows, you might even make it into the next edition of this book.

Rob Zwetsloot

FIND US ONLINE magpi.cc   **GET IN TOUCH** magpi@raspberrypi.org

EDITORIAL
Editor: **Lucy Hattersley**
Features Editor: **Rob Zwetsloot**
Book Production Editor: **Phil King**
Contributors: **Wes Archer, David Crookes, PJ Evans, Gareth Halfacree, Rosie Hattersley, Nicola King, Ben Nuttall, Marc Scott, Danny Staple, Mark Vanstone**

DISTRIBUTION
Seymour Distribution Ltd
2 East Poultry Ave, London,
EC1A 9PT | **+44 (0)207 429 4000**

DESIGN
Critical Media: **criticalmedia.co.uk**
Head of Design: **Lee Allen**
Designers: **Sam Ribbits. Mike Kay**
Illustrator: **Sam Alder**

MAGAZINE SUBSCRIPTIONS
Unit 6, The Enterprise Centre,
Kelvin Lane, Manor Royal,
Crawley, West Sussex,
RH10 9PE | **+44 (0)207 429 4000**
magpi.cc/subscribe
magpi@subscriptionhelpline.co.uk

PUBLISHING
Publishing Director: **Russell Barnes**
russell@raspberrypi.org

Advertising: **Charlotte Milligan**
charlotte.milligan@raspberrypi.org
Tel: +44 (0)7725 368887

Director of Communications: **Liz Upton**
CEO: **Eben Upton**

Contents

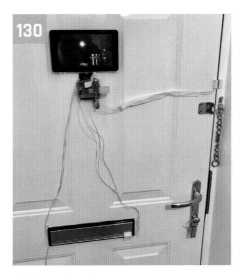

Reviews

Introducing
Raspberry

The dual-display **Raspberry Pi 4** is here to redefine personal computing

By Gareth Halfacree

A new, 28 nm system-on-chip with powerful ARM Cortex-A72 processing cores. The first new graphics processor in Raspberry Pi's history. Up to 4GB of high-speed LPDDR4 memory. Two high-bandwidth USB 3.0 ports. Dual HDMI 2.0 outputs, capable of driving a pair of 4K displays. In short: very new, very powerful, and very exciting.

Designed as a true PC replacement for a lot of use-cases, Raspberry Pi 4 is the most impressive Raspberry Pi yet – and benchmark testing proves it's far from being all talk and no substance.

Pi 4

Get to know Raspberry Pi 4

Raspberry Pi 4 marks a major Raspberry Pi family redesign

Specifications

SoC: Broadcom BCM2711B0 quad-core A72 (ARMv8-A) 64-bit @ 1.5GHz

GPU: Broadcom VideoCore VI

NETWORKING: 2.4 GHz and 5 GHz 802.11b/g/n/ac wireless LAN

RAM: 1GB, 2GB, or 4GB LPDDR4 SDRAM

BLUETOOTH: Bluetooth 5.0, Bluetooth Low Energy (BLE)

GPIO: 40-pin GPIO header, populated

STORAGE: microSD

PORTS: 2 × micro-HDMI 2.0, 3.5 mm analogue audio-video jack, 2 × USB 2.0, 2 × USB 3.0, Gigabit Ethernet, Camera Serial Interface (CSI), Display Serial Interface (DSI)

DIMENSIONS: 88 mm × 58 mm × 19.5 mm, 46 g

A | CPU
The new BCM2711B0 system–on–chip offers an impressive performance boost over its predecessors

B | POWER
The move to a USB Type-C connector for power allows Raspberry Pi 4 to support higher-current USB devices

QuickStart Guide

Raspberry Pi 4 is directly compatible with the 3B+ and all previous Raspberry Pi models, but the operating system has been updated to add support for the new system-on-chip which drives it. The easiest way to start from scratch is via our QuickStart Guide (page 18), which will guide you through setting up the latest NOOBS installer on a microSD card.
magpi.cc/quickstart

D | ETHERNET

The Ethernet port, relocated to the top-right of the board, now offers full-speed network connectivity with no bottlenecks

C | RAM

A move to up to 4GB of LPDDR4 memory, from the LPDDR2 of previous designs, increases performance further

F | USB

Two USB 3.0 ports, centre, offer high-speed connectivity for external devices including storage and accelerator hardware

E | DUAL DISPLAYS

The two micro-HDMI connectors enable Raspberry Pi 4 to drive two 4K displays

Eben Upton on Raspberry Pi 4

A brand-new processor, upgraded video capabilities, up to four times the memory – what did it take to make Raspberry Pi 4?

Eben Upton

Eben is the creator of Raspberry Pi and a co-founder of the Raspberry Pi Foundation. He is the CEO of Raspberry Pi Trading Ltd.

"**I** guess there's a question, which is 'why now, why not in a year's time?' Which is the sort of time line we previously indicated," says Eben Upton, co-founder of the Raspberry Pi Foundation, on the timing of Raspberry Pi 4's release. "Broadcom has been working on silicon for it for a little while, and the silicon came good earlier than I was expecting.

"This is the B0 step of the silicon. BCM2835, which was new on 40 nm, was equivalently radical at the time. The version we shipped there is 2835C2, so we'd had an A0, a B0, a C0, a C1, and a C2 to get to a shippable product. This one got shippable by B0, and that's taken a year out of the conservative schedule that we'd been communicating to people."

▲ Raspberry Pi 4 boards being tested and having software written and updated for them

Backwards compatibility

"It's very substantially backwards compatible," Eben promises. "You don't like to say 'perfectly backwards compatible,' because I'm sure people will find ways in which it's not. At launch, for example, I suspect there will be monitors that a Raspberry Pi 3 can drive that a Raspberry Pi 4 can't; but that will be fixed over time.

"I think we've met our goals for backwards compatibility. Which is good, because otherwise you tear your software team apart. You either have to sunset old products, which you know we hate doing, or you end up with two software teams: one to move the old product forward, and one to move the new product forward."

The pocket-sized PC

"It's a PC replacement. I mean, we've always talked about Raspberry Pi as being a PC, and that's become steadily more credible, I think, over the generations," says Eben. "I think this one takes it over the line where a lot of users will sit down in front of it and not really perceive a difference.

"You talk about the things that take you into PC land? PCs drive two displays. You know, you're not really a real PC if you don't drive two displays, right? If you think about the person you're speaking to on the phone in the bank, they'll have two monitors: one to put your account details on, and one to put the product that they're selling you on. We think this should break through very nicely into the thin client market and we're working with Citrix to make sure that their stack works on it on launch day.

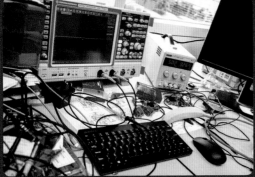

▲ A Raspberry Pi engineer's desk, where hardware is tested thoroughly and bananas are eaten

A challenging design

"Obviously the ports have moved around, and that's really a routing thing," Eben explains. "The board is within a millimetre of not working, and there wasn't enough routing resource to bring the Ethernet signalling down to the bottom-right of the board.

> ❝ The Easter egg is under the USB-C connector; it's James's signature ❞

"Probably the biggest challenge is the DRAM. If you look how close that SoC and that DRAM are to each other, you've got a 32-bit DRAM interface in that tiny little space, with some length-matching between the signals and the signals properly isolated from each other. If you were to desolder the USB-C connector, you'll see James [Adams, director of hardware] has signed the board. So, the Easter egg is under the USB-C connector; it's James's signature, because I think he feels it's the nicest piece of work he's ever done, and it was very close to not being doable.

Building the BCM2711

"This has been a more complicated development than previous ones because previous ones have been on the same process node and we've basically just been – 'just' been – bolting larger ARM cores onto an existing chip," says Eben of the work that has gone into the new system-on-chip (SoC).

"This one's on a new process node, so this one's on 28 nm. Obviously, it's got all these new features, so we've kind of moved it from being a 1080p-class chip to being a 4K-class chip. New process node, new memory technology, new multimedia IP [intellectual property], lots and lots of change. It's a full-chip project."

"The original prototypes, the A0 prototypes, are actually about five millimetres longer. They're five millimetres bigger in X than the historical board, but he was able to squeeze it back down. My contribution was largely to go to his desk and say 'is the board back at the right size yet?' every day for about six months. I should have signed the board as well – I deserve half-credit!"

Benchmarking Raspberry Pi 4

A full-chip redesign, the first in the history of Raspberry Pi, has unlocked new levels of performance

I t's not hard to see where **Raspberry Pi 4 improves on its predecessor.** The brand-new BCM2711B0 system-on-chip has more powerful processing cores, the first upgrade to the graphics processor in the history of the project, and vastly improved bandwidth for both memory and external hardware. Gone is the single-lane USB bottleneck which hampered performance on older models, and Raspberry Pi 4 shines as a result.

Spec comparison

Internally, there's little left unchanged between the Raspberry Pi 3 family and Raspberry Pi 4. The SoC is now built on a 28 nm semiconductor process node, down from 40 nm, and packs the significantly more powerful ARM Cortex-A72 processor cores. The memory has moved from LPDDR2 to LPDDR4, skipping a generation and improving bandwidth, and is for the first time available in capacities over 1GB with 2GB and 4GB versions available on launch day.

Even the graphics processor has been upgraded: the Broadcom VideoCore IV, which has been a staple since the original Raspberry Pi Model B, has been replaced with the more powerful VideoCore VI, unlocking both performance and dual-4K-display capabilities.

Add in the full-speed Gigabit Ethernet and USB 3.0 ports and you've got a significant upgrade on your hands.

Linpack

A synthetic benchmark originally developed for supercomputers, Linpack offers a glimpse at just how far the Raspberry Pi family has come. This version – ported by Roy Longbottom – comes in three variants: the fast single-precision (SP), slower double-precision (DP), and a single-precision variant accelerated using the NEON instructions available in Raspberry Pi 2 and above (NEON).

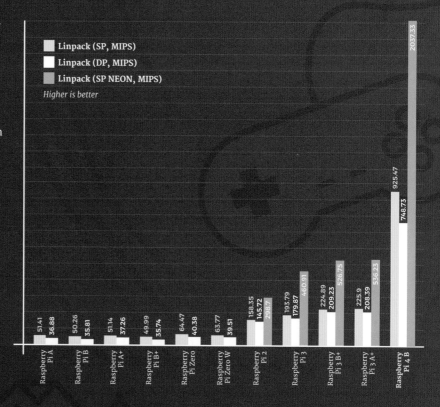

Linpack (SP, MIPS)
Linpack (DP, MIPS)
Linpack (SP NEON, MIPS)

Higher is better

	Linpack (SP, MIPS)	Linpack (DP, MIPS)	Linpack (SP NEON, MIPS)
Raspberry Pi A	51.41	36.88	
Raspberry Pi B	50.26	35.81	
Raspberry Pi A+	51.14	37.26	
Raspberry Pi B+	49.99	35.74	
Raspberry Pi Zero	64.47	40.38	
Raspberry Pi Zero W	63.77	39.51	
Raspberry Pi 2	158.35	145.72	298.7
Raspberry Pi 3	193.79	179.87	460.91
Raspberry Pi 3 B+	224.89	209.23	526.75
Raspberry Pi 3 A+	225.9	208.39	536.23
Raspberry Pi 4 B	925.47	748.73	2037.33

Python GPIO Zero

Sitting somewhere between a synthetic and a real-world benchmark, here the Python GPIO Zero library is used to toggle a GPIO pin on and off as quickly as possible while a frequency counter measures the switching rate in kilohertz (kHz). This test is boosted by CPU speed.

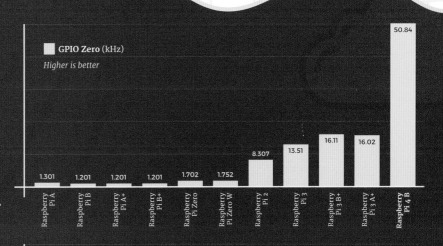

GPIO Zero (kHz)
Higher is better

Raspberry Pi A	Raspberry Pi B	Raspberry Pi A+	Raspberry Pi B+	Raspberry Pi Zero	Raspberry Pi Zero W	Raspberry Pi 2	Raspberry Pi 3	Raspberry Pi 3 B+	Raspberry Pi 3 A+	Raspberry Pi 4 B
1.301	1.201	1.201	1.201	1.702	1.752	8.307	13.51	16.11	16.02	50.84

File Compression

An example of a real-world workload, this benchmark takes a file and compresses it using the bzip2 algorithm and measures the elapsed time in seconds. For Raspberry Pi models with more than one processing core – the Raspberry Pi 2 and 3 family, and Raspberry Pi 4 – the test is run a second time using the multi-threaded lbzip2.

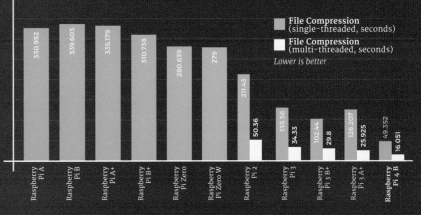

File Compression (single-threaded, seconds)
File Compression (multi-threaded, seconds)
Lower is better

Raspberry Pi A	Raspberry Pi B	Raspberry Pi A+	Raspberry Pi B+	Raspberry Pi Zero	Raspberry Pi Zero W	Raspberry Pi 2	Raspberry Pi 3	Raspberry Pi 3 B+	Raspberry Pi 3 A+	Raspberry Pi 4 B
330.952	339.603	335.179	310.738	280.639	279	211.43 / 50.36	135.58 / 34.33	102.44 / 29.8	126.207 / 25.925	49.352 / 16.051

Speedometer 2.0

Speedometer 2.0 measures the responsiveness of the Chromium web browser by running a web application – a to-do list – and measuring how many times the application can be completed each minute. Here, performance hinges not only on CPU performance but on memory speed and capacity – the test proved too much for Raspberry Pi A+.

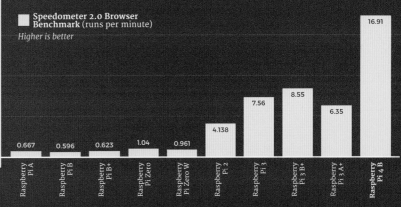

Speedometer 2.0 Browser Benchmark (runs per minute)
Higher is better

Raspberry Pi A	Raspberry Pi B	Raspberry Pi B+	Raspberry Pi Zero	Raspberry Pi Zero W	Raspberry Pi 2	Raspberry Pi 3	Raspberry Pi 3 B+	Raspberry Pi 3 A+	Raspberry Pi 4 B
0.667	0.596	0.623	1.04	0.961	4.138	7.56	8.55	6.35	16.91

OpenArena Time Demo

The new VideoCore VI gives Raspberry Pi 4 a significant boost over its predecessors, as demonstrated in this gaming workload test: the Quake III-based OpenArena first-person shooter runs through its built-in demo as quickly as possible at a High Definition (1280×720) resolution, while the average frame rate in frames per second (fps) is recorded.

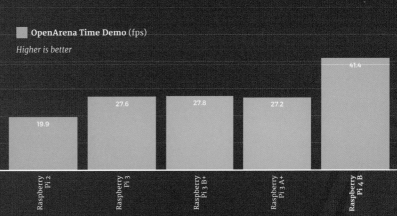

OpenArena Time Demo (fps)
Higher is better

Raspberry Pi 2	Raspberry Pi 3	Raspberry Pi 3 B+	Raspberry Pi 3 A+	Raspberry Pi 4 B
19.9	27.6	27.8	27.2	41.4

GIMP Image Editing

Another real-world test, the popular open-source GIMP image-editing suite is used to process a high-resolution image and save it as a PNG. Like the Speedometer 2.0 benchmark, this is heavily reliant on both CPU and memory performance – and extra memory really helps some of the models on test.

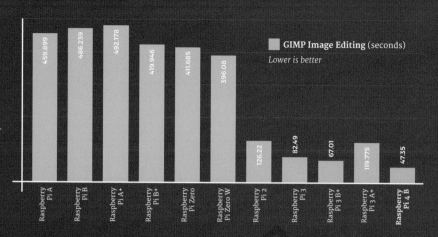

USB Storage Throughput

Raspberry Pi 4's new USB 3.0 ports offer a massive bandwidth boost, which has a big impact on the performance of external storage devices. Here, a solid-state drive (SSD) is connected via a USB adapter and the average read and write throughput measured in megabytes per second (MBps).

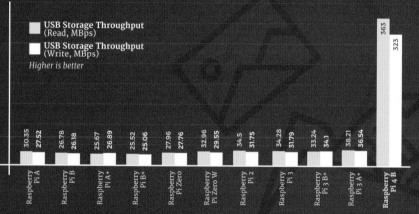

Memory Bandwidth

Although many workloads are primarily limited by CPU speed, others rely on memory bandwidth – the rate at which data can be written to and read from RAM. In this benchmark, the RAMspeed/SMP tool is used to measure the read and write bandwidth for 1MB blocks in megabytes per second (MBps).

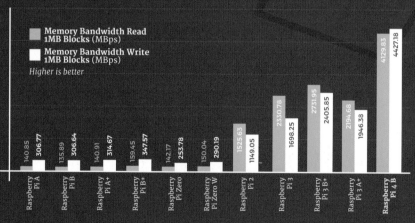

Ethernet Throughput

While Raspberry Pi 3 Model B+ added Gigabit Ethernet connectivity, throughput on Raspberry Pi 4 is free from the single shared USB 2.0 channel to the SoC. The throughput of all Raspberry Pi models with a built-in Ethernet port is measured using the iperf3 tool, showing the average network throughput (in megabits per second) over several runs.

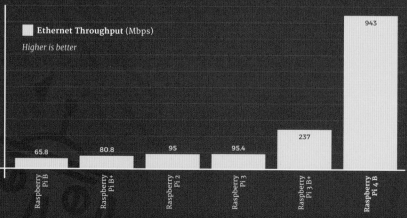

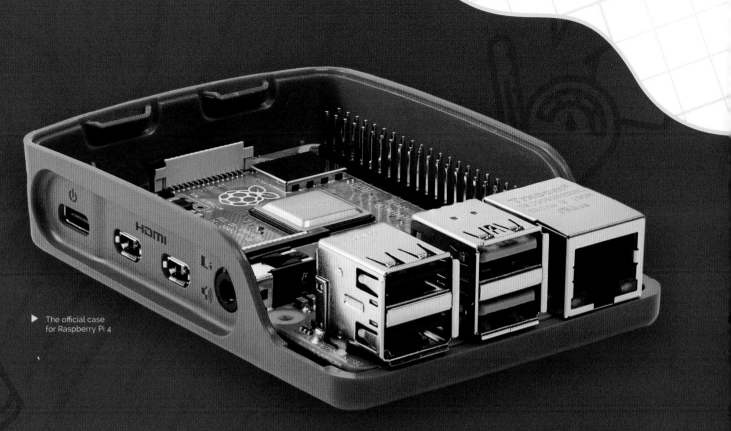

▶ The official case for Raspberry Pi 4

Wireless LAN Throughput

For this wireless networking test, an ideal environment is created: a Raspberry Pi is placed in line-of-sight of an 802.11ac router, and a wired laptop uses iperf3 to measure the average throughput over several runs. For models with dual-band 2.4 / 5GHz radios, the test is run on both bands.

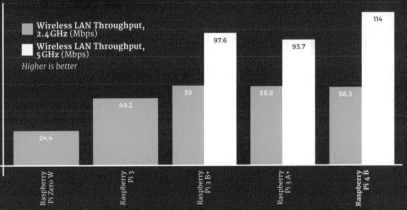

Wireless LAN Throughput, 2.4GHz (Mbps)
Wireless LAN Throughput, 5GHz (Mbps)
Higher is better

- Raspberry Pi Zero W: 24.4
- Raspberry Pi 3: 49.2
- Raspberry Pi 3 B+: 59 / 97.6
- Raspberry Pi 3 A+: 58.8 / 93.7
- Raspberry Pi 4 B: 58.3 / 114

Power Draw

More performance typically means more power, and here each Raspberry Pi model is left running a CPU-intensive benchmark while an HDMI display and a USB keyboard and mouse are connected. The peak power draw in watts is measured from the wall, and then an 'idle' draw with a Raspberry Pi sat at the Raspbian desktop is measured for comparison.

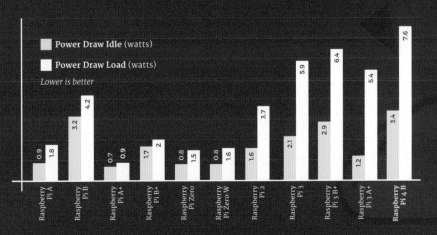

Power Draw Idle (watts)
Power Draw Load (watts)
Lower is better

- Raspberry Pi A: 0.9 / 1.8
- Raspberry Pi B: 3.2 / 4.2
- Raspberry Pi A+: 0.7 / 0.9
- Raspberry Pi B+: 1.7 / 2
- Raspberry Pi Zero: 0.8 / 1.5
- Raspberry Pi Zero W: 0.8 / 1.6
- Raspberry Pi 2: 1.6 / 3.7
- Raspberry Pi 3: 2.1 / 5.9
- Raspberry Pi 3 B+: 2.9 / 6.4
- Raspberry Pi 3 A+: 1.2 / 5.4
- Raspberry Pi 4 B: 3.4 / 7.6

Simon Long on Raspbian 'Buster'

User experience engineer **Simon Long** walks through the new features of Raspbian 'Buster' and its revamped user interface

Simon Long

Simon Long's work on user experience impacts everything you see and do on the Raspbian desktop

The launch of Raspberry Pi 4 brings not only new hardware but new software too: Raspbian 'Buster', a brand-new release – compatible, as always, with every Raspberry Pi model going right back to the pre-launch Alpha design – with a revamped, flatter user interface based on the upstream Debian 'Buster' Linux distribution.

Simon Long explains: "Due to the lack of obvious differences between Buster and Stretch, I wanted to do something to make it a bit more obvious that people actually had something new," of his new interface design. When we moved from Jessie to Stretch, there was a similar lack of major differences, and people wondered whether or not they actually had the new version – I wanted to avoid that this time. Also, the overall UI design in terms of the appearance of buttons, controls, and the like really hasn't changed significantly in the time I've been here – there have been some small tweaks, but it felt time for a change."

▲ The new Raspbian desktop offers a cleaner, more approachable interface – and lovely new wallpaper

Flatter is better

"The flatter appearance was driven by a few factors," Simon continues. "First, it does seem to be a general tendency in UI design in recent years that flatter, simpler designs are in, and fussier, more complex designs are out – iOS, Windows, and Android have all done the same sort of thing. Second, Eben is a big fan of flatter UIs, and he kept nudging me in that direction!

"It's a bit of a balancing act, though – you don't want to go too far and end up with just boring square boxes everywhere, which is why while things like corner radii have been reduced, they haven't been completely squared off.

"There's been a lot of experimentation with new designs; we toyed with things like changing the system font and considered numerous different ideas for the appearance of buttons, sliders, and scrollbars, and I think we've ended up with something that looks modern without looking too boring."

Mixing hardware and software

"**Moving to a new Debian release is always a lot of work,**" **Simon notes.** "We have to take all the changes and patches we had created for the previous version and apply them to new versions of software in the current version, test it all, make sure it is still stable and that we haven't had performance regressions, and so on.

"That on its own is usually a challenge, but the fact that we were moving to new hardware at the same time added another dimension – when you find something has broken, you don't know if it's the new hardware, the new OS, or just that you've got something wrong yourself somewhere!"

The evolution of the desktop

"**I'm really pleased with the way the new user interface design has come out,**" says Simon. "Because the design process was a gradual evolution over time, you don't realise the difference between where you started and where you've ended up, but once it was finished and I was applying the changes to existing images, the sudden switch from old to new just made everything look instantly better.

"I'd never really thought there was much wrong with the old design, but when you suddenly change to the new one, you think 'wow, that looks a lot better' – or at least I do!"

Under the hood

Not all improvements are immediately visible: "We're now using OpenGL to draw the desktop with hardware acceleration," Simon explains. "This is something which we've had as an experimental feature for a couple of years now – it's been an option in raspi-config to turn it on, but it's now the default mechanism. It means that any applications which use OpenGL should run significantly faster, and it means that things like OpenGL games are now usable on Raspberry Pi out of the box.

> " We are actually releasing Buster before Debian themselves do! "

"One interesting consequence of this is that we are actually releasing Buster before Debian themselves do! Some of the libraries which enable the OpenGL acceleration work much better in their Buster versions, so we have been using testing versions of Buster for several months and it makes more sense to release Buster software for Raspberry Pi 4 than it does to do all the work required to make this work on Stretch. Buster is in the final stages of testing by Debian – it is likely to be officially released within the next month or so – so this isn't a particularly risky thing to do, but it does mean that anyone using this release is getting it a bit early!"

Places to Buy

UK & Ireland

 Raspberry Pi Store
magpi.cc/retail-store

 The Pi Hut
thepihut.com

 OKdo
okdo.com

 CPC
cpc.farnell.com

 Pimoroni
pimoroni.com

North America

 OKdo
okdo.com

 PiShop.us
pishop.us

 Adafruit
adafruit.com

 **Newark**
newark.com

 Micro Center
microcenter.com

 Canakit
canakit.com

Europe

 BuyZero
buyzero.de

 OKdo
okdo.com

 Sertronic
digitec.ch

 Kiwi Electronics
kiwi-electronics.nl

 SEMAF
electronics.semaf.at

 Kubii
kubii.fr

 Totonic
pi-shop.ch

 Melopero
melopero.com

 Electrokit
electrokit.com

 pi3g
pi3g.com

 **Jkollerup**
raspberrypi.dk

For a full list of approved resellers please go to raspberrypi.org/products

Raspberry Pi
QuickStart Guide

Setting up Raspberry Pi is pretty straightforward.
Just follow the advice of **Rosie Hattersley**

Congratulations on becoming a **Raspberry Pi explorer.** We're sure you'll enjoy discovering a whole new world of computing and the chance to handcraft your own games, control your own robots and machines, and share your experiences with other Raspberry Pi fanatics.

Getting started won't take long: just corral all the bits and bobs on our checklist, plus perhaps a funky case. Useful extras include some headphones or speakers if you're keen on using Raspberry Pi as a media centre or gaming machine.

To get set up, simply format your microSD card, download NOOBS, and run the Raspbian installer. This guide will lead through each step. You'll find the Raspbian OS, including coding programs and office software, all available to use. After that, the world of digital making with Raspberry Pi awaits you.

What you need
**All the bits and bobs you need
to set up a Raspberry Pi computer**

A Raspberry Pi
Whether you choose a Raspberry Pi 4, 3B+, 3B, Pi Zero, Zero W, or Zero WH (or an older model of Raspberry Pi), basic setup is the same. All Raspberry Pi computers run from a microSD card, use a USB power supply, and feature the same operating systems, programs, and games.

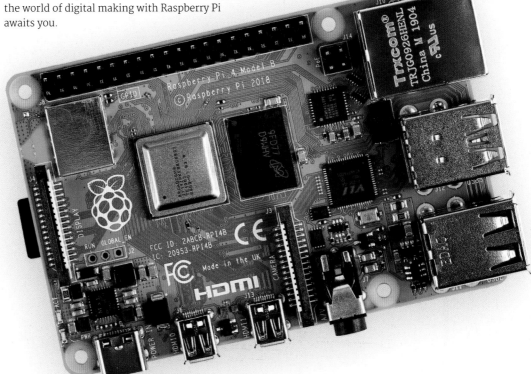

8GB microSD card

You'll need a microSD card with a capacity of 8GB or greater. Your Raspberry Pi uses it to store games, programs, and photo files and boots from your operating system, which runs from it. You'll also need a microSD card reader to connect the card to a PC, Mac, or Linux computer.

Mac or PC computer

You'll need a Windows or Linux PC, or an Apple Mac computer to format the microSD card and download the initial setup software for your Raspberry Pi. It doesn't matter what operating system this computer runs, because it's just for copying the files across.

USB keyboard

Like any computer, you need a means to enter web addresses, type commands, and otherwise control Raspberry Pi. You can use a Bluetooth keyboard, but the initial setup process is much easier with a wired keyboard. Raspberry Pi sells an official Keyboard and Hub (**magpi.cc/keyboard**).

USB mouse

A tethered mouse that physically attaches to your Raspberry Pi via a USB port is simplest and, unlike a Bluetooth version, is less likely to get lost just when you need it. Like the keyboard, we think it's best to perform the setup with a wired mouse. Raspberry Pi sells an Official Mouse (**magpi.cc/mouse**).

Power supply

Raspberry Pi uses the same type of USB power connection as your average smartphone. So you can recycle an old USB to micro USB cable (or USB Type-C for Raspberry Pi 4) and a smartphone power supply. Raspberry Pi also sells official power supplies (**magpi.cc/products**), which provide a reliable source of power.

Display and HDMI cable

A standard PC monitor is ideal, as the screen will be large enough to read comfortably. It needs to have an HDMI connection, as that's what's fitted on your Raspberry Pi board. Raspberry Pi 3B+ and 3A+ both use regular HDMI cables. Raspberry Pi 4 can power two HDMI displays, but requires a less common micro-HDMI to HDMI cable (or adapter); Raspberry Pi Zero W needs a mini HDMI to HDMI cable (or adapter).

USB hub

Instead of standard-size USB ports, Raspberry Pi Zero has a micro USB port (and usually comes with a micro USB to USB adapter). To attach a keyboard and mouse (and other items) to a Raspberry Pi Zero W or 3A+, you should get a four-port USB hub (or use a keyboard with a hub built in).

Set up
Raspberry Pi

Raspberry Pi 4 / 3B+ / 3 has plenty of connections, making it easy to set up

01 Hook up the keyboard

Connect a regular wired PC (or Mac) keyboard to one of the four larger USB A sockets on a Raspberry Pi 4 / 3B+/ 3. It doesn't matter which USB A socket you connect it to. It is possible to connect a Bluetooth keyboard, but it's much better to use a wired keyboard to start with.

02 Connect a mouse

Connect a USB wired mouse to one of the other larger USB A sockets on Raspberry Pi. As with the keyboard, it is possible to use a Bluetooth wireless mouse, but setup is much easier with a wired connection.

03 HDMI cable

Next, connect Raspberry Pi to your display using an HDMI cable. This will connect to one of the micro-HDMI sockets on the side of a Raspberry Pi 4, or full-size HDMI socket on a Raspberry Pi 3/3B+. Connect the other end of the HDMI cable to an HDMI monitor or television.

A HDMI cable, such as ones used by most modern televisions, is used to connect Raspberry Pi to a TV or display. You'll need a micro-HDMI to HDMI cable (or two) to set up a Raspberry Pi 4. Or a regular HDMI cable for Raspberry Pi 3B+ / 3 (or older) models

A regular wired mouse is connected to any of the USB A sockets. A wired keyboard is connected to another of the USB A sockets. If you have a Raspberry Pi 4, it's best to keep the faster (blue) USB 3.0 sockets free for flash drives or other components

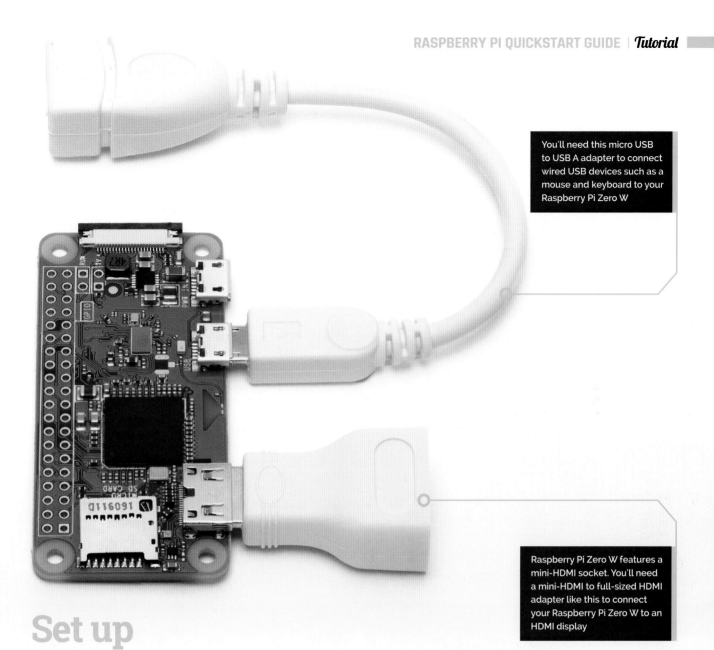

You'll need this micro USB to USB A adapter to connect wired USB devices such as a mouse and keyboard to your Raspberry Pi Zero W

Raspberry Pi Zero W features a mini-HDMI socket. You'll need a mini-HDMI to full-sized HDMI adapter like this to connect your Raspberry Pi Zero W to an HDMI display

Set up
Raspberry Pi Zero

You'll need a couple of adapters to set up a Raspberry Pi Zero / W / WH

01 Get it connected

If you're setting up a smaller Raspberry Pi Zero, you'll need to use a micro USB to USB A adapter cable to connect the keyboard to the smaller connection on a Raspberry Pi Zero W. The latter model has only a single micro USB port for connecting devices, which makes connecting both a mouse and keyboard slightly trickier than when using a larger Raspberry Pi.

02 Mouse and keyboard

You can either connect your mouse to a USB socket on your keyboard (if one is available), then connect the keyboard to the micro USB socket (via the micro USB to USB A adapter). Or, you can attach a USB hub to the micro USB to USB A adapter.

03 More connections

Now connect your full-sized HDMI cable to the mini-HDMI to HDMI adapter, and plug the adapter into the mini-HDMI port in the middle of your Raspberry Pi Zero W. Connect the other end of the HDMI cable to an HDMI monitor or television.

Set up
the software

Use NOOBS to install Raspbian OS
on your microSD card and start your
Raspberry Pi

N ow you've got all the pieces together, it's
time to install an operating system on your
Raspberry Pi, so you can start using it.

Raspbian is the official OS for Raspberry
Pi, and the easiest way to set up Raspbian on
your Raspberry Pi is to use NOOBS (New Out Of
Box Software).

If you bought a NOOBS pre-installed 16GB
microSD card (**magpi.cc/huLdtN**), you can skip
Steps 1 to 3. Otherwise, you'll need to format a
microSD card and copy the NOOBS software to it.

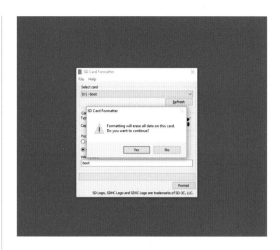

02 Format the microSD

Choose the Quick Format option and then
click Format (if using a Mac, you'll need to enter
your admin password at this point). When the card
has completed the formatting process, it's ready
for use in your Raspberry Pi. Leave the microSD
card in your computer for now and simply note the
location of your duly formatted SD card. Windows
will often assign it a hard drive letter, such as E;
on a Mac it will appear in the Devices part of a
Finder window.

01 Prepare to format

Start by downloading SD Card Formatter
tool from the SD Card Association website
(**rpf.io/sdcard**). Now attach the microSD card
to your PC or Mac computer and launch SD Card
Formatter (click Yes to allow Windows to run it).
If the card isn't automatically recognised, remove
and reattach it and click Refresh. The card should
be selected automatically (or choose the right one
from the list).

03 Download NOOBS

Download the NOOBS software from
rpf.io/downloads. NOOBS (New Out Of Box
System) provides a choice of Raspberry Pi
operating systems and installs them for you. Click
'Download zip' and save the file to your Downloads
folder. When the zip file download is complete,
double-click to launch and uncompress the folder.
You'll need to copy all the files from the NOOBS
folder to your SD card. Press **CTRL+A** (⌘+A on a
Mac) to select all the files, then drag all the files
to the SD card folder. Once they've copied across,
eject your SD card. Be careful to copy the *files inside*
the NOOBS folder to the microSD card (not the
NOOBS folder itself).

You'll Need

> A Windows/Linux
PC or Apple Mac
computer

> A microSD card
(8GB or larger)

> A microSD to
USB adapter (or
a microSD to
SD adapter and
SD card slot on
your computer)

> SD Memory Card
Formatter
rpf.io/sdcard

> NOOBS
rpf.io/downloads

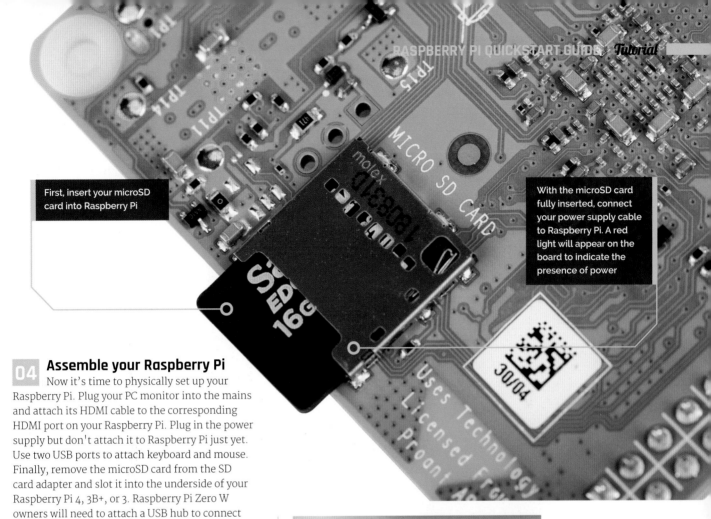

First, insert your microSD card into Raspberry Pi

With the microSD card fully inserted, connect your power supply cable to Raspberry Pi. A red light will appear on the board to indicate the presence of power

04 Assemble your Raspberry Pi

Now it's time to physically set up your Raspberry Pi. Plug your PC monitor into the mains and attach its HDMI cable to the corresponding HDMI port on your Raspberry Pi. Plug in the power supply but don't attach it to Raspberry Pi just yet. Use two USB ports to attach keyboard and mouse. Finally, remove the microSD card from the SD card adapter and slot it into the underside of your Raspberry Pi 4, 3B+, or 3. Raspberry Pi Zero W owners will need to attach a USB hub to connect mouse, keyboard, and monitor; the microSD card slot is on the top of its circuit board.

05 Power up

Plug in your Raspberry Pi power supply and, after a few seconds, the screen should come on. When the NOOBS installer appears, you'll see a choice of operating systems. We're going to install Raspbian, the first and most popular one. Tick this option and click Install, then click Yes to confirm. For more OS options, instead click 'Wifi networks' and enter your wireless password; more OS choices will appear. Installation takes its time but will complete – eventually. After this, a message confirming the success installation appears. Your Raspberry Pi will prompt you to click OK, after which it will reboot and load the Raspbian OS.

06 Get online

When Raspbian loads for the first time, you need to set a few preferences. Click Next, when prompted, then select your time zone and preferred language and create a login password. You're now ready to get online. Choose your WiFi network and type any required password. Once connected, click Next to allow Raspbian to check for any OS updates. When it's done so, it may ask to reboot so the updates can be applied.

Click the Raspberry icon at the top-left of the screen to access items such as programming IDEs, a web browser, media player, image viewer, games, and accessories such as a calculator, file manager, and text editor. You're all set to start enjoying your very own Raspberry Pi. 🅼

RASPBERRY PI 4

STARTER GUIDE

Get to know your brand new Raspberry Pi 4

By **Sean McManus**

E ver since the launch of Raspberry Pi 4, we've been seeing a lot of people on social media get stuck in with their new incredible computer. We've also seen a lot of new Raspberry Pi users finally take the plunge and get their first Raspberry Pi, and to those people we say: welcome!

Some of you may still be getting your heads around your new Raspberry Pi, so we're here to help with our Starter Guide that should teach you some of the basics (and beyond!) of how to use it. Grab a microSD card and find a spare monitor, because it's time to have some fun.

Basic setup

01 The heart of your new computing experience: Raspberry Pi 4. Find out the full, amazing specs here: **magpi.cc/benchmarks**

02 Power up to two 4K monitors with Raspberry Pi 4's dual micro-HDMI ports

03 With a keyboard and mouse, you can easily use it as a desktop computer

04 The USB 3.0 ports allow for high-speed file transfers

FIRST STEPS WITH RASPBERRY PI

Now you've set up your Raspberry Pi, discover some of its accessories and explore the Raspbian operating system. Sean McManus is your guide

Raspberry Pi Case

Protect your Raspberry Pi from spills and dust with a chic case. Official Raspberry Pi cases come in red/white and grey/black, but there are plenty of cases available from other companies too. If you'll be building electronics projects, look for a case that gives you easy access to the GPIO.

magpi.cc/YNvYfF

Sense HAT

HATs (short for Hardware Attached on Top) are accessories that plug onto a Raspberry Pi's GPIO pins. The Sense HAT includes a colourful 8×8 grid of LEDs and a five-button joystick. It's packed with sensors: gyroscope, accelerometer, barometric pressure sensor, magnetometer, thermometer, and humidity sensor. It's a portable science lab!

magpi.cc/sense-hat

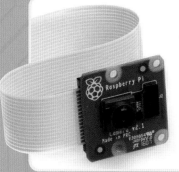

Raspberry Pi Camera Module V2

You can plug a Raspberry Pi Camera Module into a dedicated connector on Raspberry Pi and it has an 8-megapixel sensor. It works with Raspberry Pi 1, 2, 3, and 4. Shoot a movie in high definition, get close to nature with a bird-box cam, or secure your home.

magpi.cc/camera

Fan SHIM

Raspberry Pi 4 is a bit more power hungry than earlier models. If you find it runs hot for your application, try using the Fan SHIM. It provides a software-controllable fan to cool a Raspberry Pi. You can use it together with HATs, by fitting a booster header to lift them above the fan. Check out our review on page 188.

magpi.cc/qZYBWd

CamJam EduKit #3

With its easy control of electronic circuits, Raspberry Pi is ideal for robots. This kit contains everything needed to build your first robot, including a motor controller board, sensors, and wheels. Make a chassis from Lego, 3D-print one, or even use the box.

magpi.cc/RhpjZh

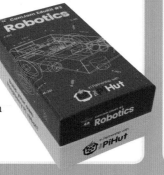

Meet Raspbian with Desktop

01 Click the Raspberry Pi logo to open the applications menu. This is where you'll find the software that's pre-installed in Raspbian, and anything else you add later.

02 Click the globe for quick access to the Chromium web browser.

03 Click the folders to open the File Manager. You can use it to find, move, copy, and delete files on your storage devices. Why not explore the Linux file system?

04 Find the Terminal here. It gives you powerful tools for managing your files and devices, and the command line is often the quickest way to get things done.

05 Volume control. Right-click to select audio output.

06 WiFi options. Turn WiFi on and off, and switch networks here. If the icon is blue like this, you're connected. You can also

hover the mouse pointer over the icon to see your Raspberry Pi's IP address.

07 Manage Bluetooth connections. You can use Bluetooth devices such as keyboards and mice to

wirelessly control your Raspberry Pi device.

08 File Manager. Use the hierarchical browser on the left or the Go menu to find devices connected to Raspberry Pi. You should

store your files in the **/home/pi** folder.

09 With basic, scientific, and paper modes, this calculator is handy. Find it in the Accessories section of the applications menu.

Raspberry Pi Configuration

Need to adjust some settings? The Raspberry Pi Configuration tool is in the Preferences section of the applications menu. In its System tab, you can change your password, adjust display options, and set Raspberry Pi to boot to the command-line interface (CLI) instead of the desktop.

In the Interfaces tab, you can enable connections, including remote GPIO access and the camera.

To adjust the amount of memory for the GPU or to set older Raspberry Pi models to run faster (overclocking), visit the Performance tab.

You can change the time zone, keyboard, and other geographic options in the Localisation tab.

There are separate options in Preferences for configuring the appearance, audio, main menu, mouse and keyboard, and screen.

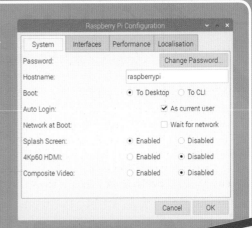

AWESOME APPLICATIONS

Discover some of the software that comes with Raspbian, and find out how to install more

Raspbian comes with a selection of pre-installed software (which depends on which version you install), so you can start working, learning, and making things straight away. Ranging from productivity suites to games, you'll find a well-curated collection of software is just a click away, in the applications menu. There are lots more packages to browse and install, too.

LibreOffice Writer | OFFICE

No computer is complete without a word processor. LibreOffice Writer has all the font and formatting options you would expect, and has basic compatibility with Microsoft Word.

LibreOffice Calc | OFFICE

Work out your budgets with this spreadsheet package. If you're familiar with Microsoft Excel, you'll feel at home here. LibreOffice Calc can open and use typical Excel files.

LibreOffice Impress | OFFICE

If you're presenting to a room of people, don't panic: Impress has your back. Use it to craft and display your slide deck. It's largely compatible with Microsoft PowerPoint.

Chromium | INTERNET

Chromium is the open-source version of Google Chrome. The default search engine in Raspbian is Duck Duck Go, which promises not to track you online.

Claws Mail | INTERNET

Send and receive email. The setup wizard helps you add your email account, and the streamlined interface shows you your mailboxes, message list, and a preview of the selected message.

VLC | SOUND & VIDEO

Listen to music while you're coding using VLC, a fully featured media player for music and video. It can play digital files, streams, and physical media such as CDs and DVDs.

Minecraft Pi | GAMES

We call it a game, but it's more a way of life. With Minecraft Pi, you can build things in Creative mode, and write programs to change the (game) world.

Python Games | GAMES

These tea-break games are fun, but they're also great Python demos. Find the code in the **/usr/share/python_games** folder, and open it in a Python editor to see how it works.

SmartSim | PROGRAMMING

Experiment with designing and testing digital logic circuits. You can develop custom components, and then incorporate them into other circuits. Download examples and read tutorials at **smartsim.org.uk**.

Sense HAT Emulator | PROGRAMMING

This emulator features on-screen controls to simulate temperature, pressure, humidity, and device position changes. A great way to try out the Sense HAT before you buy one!

Updating your software

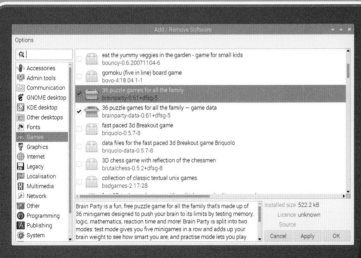

Linux software comes in packages, which are compressed archives that you can download. A package manager is used to find and install them, including any other software they need to work.

Use Add/Remove Software to update your packages. From the menu, choose Preferences > Add / Remove Software. Click on Options and Check for Updates.

You can also update from Terminal. In Raspbian, the package manager is called APT. Updating all the software from the Terminal is a two-step process. First, enter `sudo apt update` in the Terminal to update the cache of available software. Then enter `sudo apt upgrade` to update the software installed on Raspberry Pi. You'll be told what changes will be made and asked to confirm by typing **Y** and pressing **ENTER**.

Why not install these?

Fritzing | PROGRAMMING
Design and document your electronics circuits based on Raspberry Pi with this design tool. We use it to make circuit diagrams in *The MagPi*.

GNU Image Manipulation Program (GIMP) | GRAPHICS
GIMP is a powerful image editor. You can use it to create digital art, but its best feature is probably the clone tool, which enables you to retouch holiday photos.

Mathematica
PROGRAMMING
Good for more than just maths, Mathematica uses the Wolfram language, which has data and intelligence built in. Install it from Recommended Software, in the Preferences section of the applications menu.

Installing software

The easiest way to manage software is to use the Add / Remove Software tool in the Preferences part of the applications menu. It provides a visual front-end for the package manager.

You can click a category on the left to browse applications, or enter a keyword in the search box in the top-left to look for a particular application. Choose the applications you want to install by ticking the box beside them. Some packages require other packages to work properly, but the tool will take care of that for you. To remove an application, untick its box. When you've made your choices, click OK to install or remove your software.

GET CONNECTED

If your Raspberry Pi is in a tree photographing nature, you don't want to climb up there just to update it. Log in remotely and take control

There are two technologies you can use to connect to your Raspberry Pi: Secure Shell (SSH) and Virtual Network Computing (VNC). Before you can use them, they need to be switched on in the Raspberry Pi Configuration settings. To do so, click the Interfaces tab, then enable SSH and/or VNC.

Both approaches require you to know the IP address of your Raspberry Pi device. To find out, click the Terminal icon on the taskbar, and enter `ifconfig` at the prompt. It will show you all your network connections. You're looking for an IP address, which will be four numbers with a dot between them, like 198.51.100.0. You'll find it beside 'inet', in the details for wlano if you're using WiFi or in the etho summary if you're plugged in to the network.

Using SSH

SSH enables you to use the command line on your Raspberry Pi remotely, so it's very handy for installing software and fixing configuration files. You can't use SSH to run any applications that need the graphical desktop, though.

The software for SSH is pre-installed on Linux, macOS, and Windows 10. Start by opening the command line. On a Mac, find it by typing 'Terminal' into the Spotlight search. On Windows 10, Use ⊞+R to open the Run dialog box, then type in **cmd**.

Once you're in the command line, enter `ssh pi@198.51.100.0`, but replace the numbers with the IP address of your own Raspberry Pi device. The 'pi' bit is your username.

> ❝ SSH enables you to use the command line on your Raspberry Pi remotely, so it's very handy ❞

The first time you connect to a device using SSH, you'll see a warning that shows the ECDSA key for the device you're trying to connect to. You can (if you wish) validate that this is correct by using `ssh-keygen -l -f /etc/ssh/ssh_host_ecdsa_key.pub` in the Raspberry Pi Terminal. But it's usually OK to just type in **yes** and then press **ENTER** to confirm you want to connect.

You'll be asked to enter the password for your Raspberry Pi device. You won't see the cursor move while you do this, so type on regardless. When you press **ENTER**, you're in! You'll see a Linux welcome message, the date and time of your last login, and then the Raspbian command prompt, waiting for instructions. When you've finished, enter `exit` to leave the SSH session.

If you're using an earlier version of Windows, download PuTTY from **putty.org**. Enter your Raspberry Pi's IP address in the Host Name box, and click Open. Again, you'll see a warning if this is your first connection, which you can safely dismiss. Log in as **pi**, and enter your password. Raspberry Pi is now at your command!

Tip!

See our SSH tutorial on page 40 for more detailed information on using SSH to remotely control a Raspberry Pi.

Raspberry Pi Configuration			
System	Interfaces	Performance	Localisation
Camera:	○ Enabled	● Disabled	
SSH:	● Enabled	○ Disabled	
VNC:	● Enabled	○ Disabled	
SPI:	○ Enabled	● Disabled	
I2C:	○ Enabled	● Disabled	
Serial Port:	○ Enabled	● Disabled	
Serial Console:	● Enabled	○ Disabled	
1-Wire:	○ Enabled	● Disabled	
Remote GPIO:	○ Enabled	● Disabled	
		Cancel	OK

▲ Enable SSH and VNC before you try to use them

▲ Using VNC Viewer to manage Raspberry Pi from an iPad

Using VNC

Virtual Network Computing (VNC) enables you to remotely access the Raspberry Pi desktop, so you can manage files and run software using it. Some people use VNC to share their keyboard, mouse, and monitor between a PC and Raspberry Pi.

Raspbian includes VNC Server, which runs automatically in the background if you've enabled it in your settings. You'll need to download and install VNC Viewer (**magpi.cc/FuGnye**) on the device you want to use to control Raspberry Pi. VNC Viewer is available for Windows, macOS, and Linux. There are also Android and iOS apps, so you can use VNC to control Raspberry Pi from a mobile device, although it's rather fiddly without a real mouse and keyboard. Optionally, by creating a RealVNC account and registering your Raspberry Pi, you can then access it from anywhere in the world using VNC Viewer.

When you start VNC Viewer, it'll ask you to enter a VNC Server address. This is the IP address you noted when you ran `ifconfig` on your Raspberry Pi.

The first time you connect to a device, VNC Viewer warns you that it has no record of connecting to this device before, and shows you the device's signature and identification catchphrase. To verify you're connecting to the correct device, click the VNC icon on the right of the taskbar in Raspbian to see your device's details.

VNC Viewer prompts you to enter the username and password for the device you're connecting to. You'll then see your Raspbian desktop in the VNC Viewer window. You can now use your keyboard and mouse (or touchscreen on a mobile device) to control the Raspbian desktop, including using programs installed on Raspberry Pi.

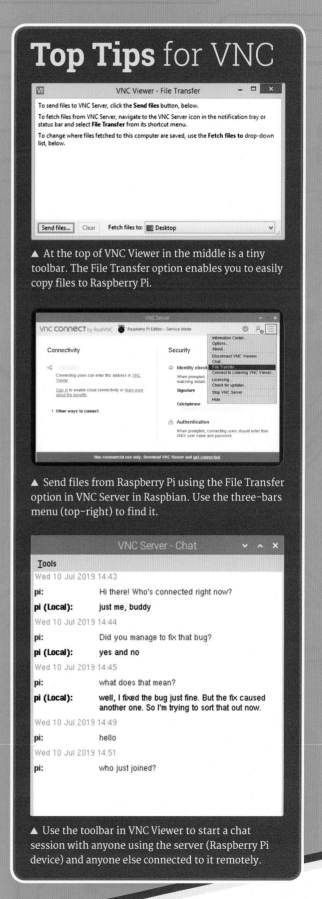

Top Tips for VNC

▲ At the top of VNC Viewer in the middle is a tiny toolbar. The File Transfer option enables you to easily copy files to Raspberry Pi.

▲ Send files from Raspberry Pi using the File Transfer option in VNC Server in Raspbian. Use the three-bars menu (top-right) to find it.

▲ Use the toolbar in VNC Viewer to start a chat session with anyone using the server (Raspberry Pi device) and anyone else connected to it remotely.

CODE TO JOY

Raspberry Pi comes with everything you need to start programming

If you've never programmed before, you're in for a treat. Raspbian comes with several integrated development environments (IDEs) you can use to write your own programs, supporting languages that are friendly to use, and fun to tinker with. If you're a veteran coder, you'll find powerful IDEs are included to help you be productive, too.

Scratch 3
LANGUAGE: SCRATCH

With its friendly drag-and-drop commands and a library of sprites and sound effects, Scratch makes it simple to get started with coding. By minimising typing, and guiding you to sensible block combinations, it helps you to avoid common mistakes. Don't be fooled by its accessibility, though: it's a fully-fledged programming language, with plenty of potential.

Thonny
LANGUAGE: PYTHON

Python is one of the most popular languages on Raspberry Pi, and Thonny is our favourite way to edit it. In a single window it shows you your code, the shell, and your data (variables), so you can easily see what's going on and fix any bugs that creep in.

Sonic Pi
LANGUAGE: SONIC PI

Learn how to compose and perform music with code! Sonic Pi incorporates synths, samples, and effects that you can control by writing programs using a language based on Ruby. You can develop and modify programs while the music plays, to perform live concerts or improvise at home.

Node-RED
LANGUAGE: JAVASCRIPT/NODE-RED

Node-RED enables you to use a flowchart to program data flows for Internet of Things applications on Raspberry Pi. It uses JavaScript, the language of the web. Run the Node-RED console in Raspbian and then visit **http://localhost:1880** in your browser to program it. For help, see **nodered.org**.

Greenfoot
LANGUAGE: JAVA

Java is one of the world's most popular programming languages. Greenfoot makes it easier to learn, by providing a friendly environment for building simple games. The editor incorporates a game world, and you add Java code to image objects to control their interactions. Find out more at **greenfoot.org**.

Geany
LANGUAGE: LOTS!

If you're looking for a lightweight but powerful IDE, try Geany. It supports many languages, including HTML, C, Java, PHP, JavaScript, and Perl. Its code auto-completion and syntax highlighting can help you code faster, and reduce errors. Our favourite feature? Code folding, used to show or hide logical chunks of code in a long program.

FURTHER RESOURCES

There's a wealth of resources available to support you as you learn more about Raspberry Pi and Raspbian

The Official Raspberry Pi Beginner's Guide

Available to buy in print and as a free PDF, this book shows you how to set up Raspberry Pi, and gets you started with programming it in Scratch and Python. It also covers the Sense HAT and Raspberry Pi Camera Module, with code examples you can build and tailor.

magpi.cc/BGbook

Official Raspberry Pi Documentation

The documentation provides concise user guides for Linux, Minecraft, Sonic Pi, Scratch, and Python. It's also the place to go for advice on configuring Raspberry Pi, hardware specifications, and remote access tips. Useful for both beginners and power users.

rpf.io/docs

GPIO Zero Documentation

See how easy it is to start programming your own electronics projects for Raspberry Pi. The documentation for GPIO Zero shows you how to connect up sensors, LEDs, motors, and more. With a bag of cheap components, you can start building your first circuits.

rpf.io/gpiozero

Conquer the Command Line

We publish a series of short books called *The MagPi Essentials*, and this one explains the Terminal, including using it for connecting disks, compiling software, and backing up. You can download all the books in the series, and past issues of *The MagPi*, at **magpi.cc/issues**.

magpi.cc/CLIbook

Hacking and Making in Minecraft

Minecraft on Raspberry Pi is a great way into the world of coding. This book, another in *The MagPi Essentials* series, shows you how to shape the world with Python code, use a connected LED as a treasure sensor, and use Sonic Pi and Node-RED together with Minecraft.

magpi.cc/MCbook

Python Games

Go behind the scenes of the Python games that come pre-installed in Raspbian with Al Sweigart's free online book. It explains the program code for seven of the games in detail, and lets you study another four listings yourself.

magpi.cc/RTfkvD

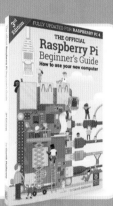

Raspbian Buster's desktop interface will be familiar and it features a welcome wizard

Raspberry Pi 4 comes in three RAM sizes – we tested the flagship 4GB version

You can use any standard HDMI monitor, or even set up dual screens

Any USB (or Bluetooth) keyboard may be used. The official one even has a three-port USB hub

A standard USB (or Bluetooth) mouse may be used. An official one is included in the Desktop Kit

RASPBERRY PI 4

YOUR NEXT
DESKTOP
PC

PJ Evans puts a Raspberry Pi 4 to the ultimate test: a full desktop replacement

When Raspberry Pi 4 was launched earlier in 2019, the significant improvements in processor speed, data throughput, and graphics handling lead to an interesting change of direction for this once-humble small computer. Although impressive that you can run a 'full' Linux operating system on a $35 device, a lot of people were just using it to get Scratch or Python IDLE up and running. Many people were skipping the graphical side altogether, and using smaller models, such as Raspberry Pi Zero, for projects previously covered by Arduino and other microcontrollers.

Raspberry Pi 4 was different. Tellingly, the Raspberry Pi Foundation released a new all-in-one kit and named it the 'Desktop Starter Kit'. For the first time truly in Raspberry Pi history, it was considered powerful enough to be used as a daily computer without any significant compromise. Challenge accepted. We asked PJ Evans to spend a week using a Raspberry Pi 4 as his only machine. Here's what happened.

Day 1 | **Monday**
Decisions, decisions

Our new favourite single-board computer comes in a selection of RAM sizes: 1GB, 2GB, or 4GB. With a difference of £20 between the 1GB and 4GB versions, it made sense to go right for the top specification. That's the version included in the official Desktop Kit (**magpi.cc/hDYcvr**) that I went out and bought for £105 (inc. VAT) at the official Raspberry Pi store – it normally retails for $120 plus local taxes. My last laptop was £1900. I'm not suggesting that the two can be reasonably compared in terms of performance, but £1795 minus the cost of a monitor is a difference worth remarking upon.

Back at the office, I inspected the contents. For your money you get a 4GB version of Raspberry Pi 4, thoughtfully already installed in the new official case; the official keyboard and mouse; the new USB-C power supply; a 16GB microSD card preloaded with the Raspbian Buster operating system; and a copy of *The Official Raspberry Pi Beginner's Guide* 252-page book. It's very well packaged and presented, with little plastic waste. The book is the icing on the cake if you are looking at this set for a young person's first computer, short-circuiting the 'now what do I do?' stage. What pleased me

▼ The Desktop Starter Kit contains everything you need to get started

in particular was the inclusion of two micro-HDMI cables in the kit, allowing me to set up a dual-screen system without delay.

First tests

I set up my new workstation next to my existing laptop, with two 1080p monitors that only had DVI connectors, so I had to get a couple of £2 adapters (**magpi.cc/NwxNFb**) and an additional cable to get sound out of the audio jack of my Raspberry Pi. Time for an initial test-drive. Booting up into Raspbian Buster was quick, about ten seconds, and connection to WiFi easy. There's no doubting the feel of the speed improvements. Yes, I've read all the benchmark tests (**magpi.cc/benchmarks**), but I wanted to know how that translates to user experience. This new kit does not disappoint.

Raspbian has matured impressively as an OS. For my daily desktop scenario, the jewel in the crown is Chromium: having such a capable web browser is what makes this whole experiment feasible. Others have upped their game, too: Firefox has come a long way, and many other browsers are now available, such as Vivaldi (**vivaldi.com**). A check of some of my most visited sites showed Chromium to be just as capable as Chrome on my regular machine. Unsurprisingly, it wasn't as snappy and I hit a few bumps, but we'll get to that.

A day of impressions

I'm no expert when it comes to GPUs, but I was impressed with the dual-monitor support. The setup worked first time and didn't seem to have any detrimental effect on the machine's performance. I was expecting slow window drawing or things getting 'stuck', but this wasn't the case.

By the end of the first day, I was getting used to the keyboard and mouse too. They are a nice mixture of being both functional and aesthetically pleasing. The keyboard comes with a three-port hub, so you can connect the mouse if you wish. It does not have the build quality and precision of my daily wireless keyboard and trackpad, but for a fraction of the price I was surprised how much you got for your money. By the end of the week I'd grown quite fond of it.

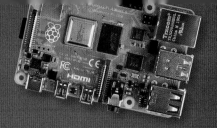

Day 2 | **Tuesday**
Back to basics

here's a much shorter version of this feature that reads 'Install Raspbian, use web apps, it'll be great'. Seriously, if you're OK with the cloud, you could now go for a long swim in the Google App waters and have nothing more to worry about. Gmail, Calendar, Drive, Docs, Spreadsheet, and the list goes on (see **gsuite.google.com**). A full suite of business software awaits you, and the same can be said for Office365 (**office.com**). I successfully used both and although they do hit the CPU hard, both were functional and certainly the most friction-free way of doing day-to-day work. However, many of us would prefer to use open-source software, have privacy concerns, or do not wish to rely on a good network connection, so let's look at the alternatives.

When it comes to the standard suite of office applications – word processing, spreadsheets, and presentation software – Raspbian users are already set up and good to go with the LibreOffice suite. Six applications, all for free and, in their current incarnations, more than fit for purpose. Previous Raspberry Pi models have struggled to run these admittedly large apps at speed, but Raspberry Pi 4 has no such issues. In fact, this feature has been written solely using LibreOffice Writer and although it's noticeable that you're not on the world's fastest computer, it is more than adequate for even complex word processing and layout.

> ❝ Users are already set up and good to go with the LibreOffice suite ❞

Beyond installed software

I'm going to need email, calendars, and contacts. I ideally want to be able to access these anywhere. My current email provider uses IMAP, so that gave me a choice of clients to look at. The Claws email client is provided with Raspbian. Setup went smoothly but although fast, I found it lacking in features and had an old-school appearance that

▲ The desktop setup in all its dual-monitor glory

reminded me of Eudora from the 1990s. Luckily, an old favourite, Thunderbird, was available to install (**thunderbird.net**). This long-term Mozilla project is still very popular. Installation and configuration was straightforward. I couldn't say the same for calendaring. Lightning, the calendar plug-in for Thunderbird, refused to talk to my remote iCloud calendar. Some research showed that Apple, although supporting the CalDAV protocol, doesn't play nice and several clients have issues.

In the end, my platform of choice for email, calendars, and contacts was Evolution (**magpi.cc/cXJvYJ**). This attractive, fast, and actively maintained app provided everything I needed. It's not installed by default, but can be swiftly installed using 'Add/Remove Software' under 'Preferences' in the Raspbian menu. If I was happy with a local calendar and contact list, that would be that, but I'd prefer to have access on my phone too, wherever I am. Currently, I use iCloud to do this, but is there a workable open-source alternative where I own the data?

How to install?

Most of the software featured can be installed using 'Add/Remove Programs' in Raspbian's desktop menu, or APT on the command line.

Day 3 | **Wednesday**
Into the cloud

In for a penny, I thought, and decided to look for an alternative to my current cloud provider that would support all the common protocols and allow me to keep control over my data. After a few false starts, two packages provided workable solutions.

Nextcloud (**nextcloud.com**), a fork of popular open-source solution ownCloud, provides file storage, calendaring, contacts, note taking, and task management. Its well-supported plug-in system allows you to extend those capabilities further. Raspberry Pi is normally seen as a server platform for Nextcloud, but now we're looking to be the client too. To my surprise, I was able to get the Nextcloud server running easily on my Raspberry Pi 4 using

> ❝ I now had a locally running web service that offered network access to calendars, contacts, and more ❞

Docker, a system for 'containerising' complex software by wrapping it in a cut-down operating system. I now had a locally running web service that offered network access to calendars, contacts, and more. Evolution connected first time, supporting tasks as well. Nextcloud offers apps for iOS and Android, allowing you to sync calendars, contacts, and files across platforms.

Cloud saviour

Nextcloud's Files feature is particularly worthy of note. This can replace services such as Dropbox in your workflow by acting as a central file repository. Your Raspbian desktop can connect using WebDAV, a file networking protocol, but if you're after proper file sync, there is a Nextcloud Desktop application available. Unfortunately, it's not available from repositories yet, but instructions are on Nextcloud's forums on how to get the source code and compile it yourself. I now have a folder in my home directory that is always in sync with the server. Best of all, if I go to the server's built-in website, I can share files and create one-off links that I can give to others.

The only hitch is where to run the server. For testing, I was running it locally (and it's worth pointing out that my Raspberry Pi was quite happy running my desktop, Docker, and the NextCloudPi (**magpi.cc/spLCMp**) server image: near-zero CPU while idling and only 2GB of memory in use), but that means outside of my local network I would have no access. In the long-term, I'm going to put Nextcloud on a dedicated Raspberry Pi and configure port-forwarding on my router. Alternatively, I may set up a virtual server 'in the cloud' so I can run it from there. A complete file and data sharing solution and I own all the data.

Syncthing (**syncthing.net**) is a simpler way of getting files about without worrying about servers. Just install Syncthing on every machine with which you want to share files. On each machine, you can select which directories you want to share and with whom. It's very easy to set up, detecting other Syncthing instances on the local network and quickly linking them up if you so wish. It's also highly configurable, with features such as selective syncing and bandwidth throttling. Syncthing is available for a wide range of platforms, so this is a great way to collaborate with Windows and macOS users. However, there are currently no apps for mobile devices available, so access to files 'on the go' is restricted.

Honorary mention: Rclone (**rclone.org**) – an open-source command-line tool for syncing directories with a number of third-party services such as Dropbox and Google Drive.

Day 4 | **Thursday**
Photos and video

Now for something a bit trickier. Raspberry Pi 4 comes with a dedicated GPU, the Broadcom VideoCore VI, making it by far the most powerful Raspberry Pi in terms of graphics performance. This is great news, but is it good enough to cover the demands of photo and video work? Surely the larger PC cousins of Raspberry Pi will have the edge?

The first 'real world' problem to be solved was how to get photos onto Raspberry Pi in the first place. I have an iPhone, and iOS is notoriously uncooperative with things that are not Apple. I could fumble around with forwarding photos via email or using Rclone with Dropbox, but I wanted something far more elegant like the workflow I was used to: take the photo and it appears on your computer. Thankfully, Nextcloud came to the rescue again. The iOS and Android apps can both automatically sync photos back to the server as they are taken. The Nextcloud Desktop app in turn syncs the new files to the local file system. All tests worked perfectly.

First attempt in learning
At this point, I hit the first failure of this experiment. The people's choice for Linux photo management software appeared to be digiKam (**digikam.org**), and the features and screenshots did look impressive. Despite several attempts, however, the software flatly refused to run, getting stuck when starting up. This may be due to my using Raspbian Buster, or some other issue I was unable to identify. Instead I settled on Shotwell (**magpi.cc/uLYpzW**), which not only worked first time but provided an intuitive and simple interface. Shotwell provides all the basic functionality you need to manage a photo library and provides the fundamentals for making small touch-ups and corrections to images, such as auto-enhance, cropping, rotation, and colour adjustments.

For more comprehensive photo editing, GIMP (**gimp.org**) is probably the most popular choice in the open-source community: a completely free, full-featured smorgasbord of image editing tools matched only by professional packages like Adobe Photoshop. This is a beast of an application and has struggled on previous Raspberry Pi models. It's still not the fastest experience, and would quickly become frustrating if you were working in it all day every day, but for the occasional bit of work it is perfectly usable. If GIMP's complexity looks like too much for you, check out Mirage instead.

Video editing was always going to be a big ask for such a low-cost platform. I installed OpenShot (**openshot.org**), a non-linear multitrack video editing platform and, sure enough, although it worked, it did struggle with the sample 720p MP4 video I imported. Raspberry Pi has dedicated hardware to handle the H.265 (HEVC) video codec only, so anything else has to be processed solely by software.

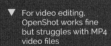

▼ For video editing, OpenShot works fine but struggles with MP4 video files

▲ The GIMP image editor is a sophisticated application and runs well enough on Raspberry Pi 4

Day 5 | **Friday**
Fun

Which OS?

We've focused on Raspbian here, but you may consider Ubuntu MATE, CentOS, or openSUSE. Get multiple microSD cards and try them all!

So, the working week is almost done and it's time to relax. How can my little desktop cope with gaming? It's fair to assume I won't be playing Fortnite or F1 2019 at 60 fps on Raspberry Pi any time soon, although the ubiquitous Minecraft works very well. Luckily for me, I enjoy retro gaming and Raspberry Pi is more than capable of running various emulators for 1980s and 1990s game consoles. RetroPie (**retropie.org.uk**) is a wonderful platform that brings multiple emulators together in a common interface, greatly simplifying installation and configuration (Note: At the time of writing, support was still being finalised for Raspberry Pi 4). If you're after something a bit more recent, there are many casual games out there, including ports of early PC classics like Doom. You can certainly be entertained by your Raspberry Pi desktop, but don't expect a VR headset experience.

How about movies, music, and podcasts instead? Raspberry Pi's potential as an affordable media streaming platform has been much vaunted. If you want a full media centre experience, you can choose between different implementations of the Kodi platform (**kodi.tv**), such as OSMC (**osmc.tv**), LibreELEC (**libreelec.tv**), and more. These typically come as full disk images

▲ Plex is a good choice for streaming videos, music, and podcasts from your networked-attached storage

and are not intended to be run from the desktop. Although Kodi can be installed on the desktop just like digiKam, I struggled to get it to run. For these kinds of applications, I would seriously consider using multiple microSD cards. One of the joys of Raspberry Pi is you can easily swap out cards and, presto, your desktop machine is now an arcade or a media centre.

> ❝ You can certainly be entertained by your Raspberry Pi desktop ❞

Watching YouTube videos was hit and miss. Cinema mode worked well, but full-screen struggled and we suffered a few crashes. I had much better results with Plex streaming from my local server. A 720p video played flawlessly, although some tearing was in evidence. If you want to access other streaming services such as Netflix, Kodi, and friends are the way to go. Chromium does not support the DRM (digital rights management) required by Netflix.

Day 6 & 7 | **Weekend**
Conclusions

There is no doubt, Raspberry Pi is now a capable desktop computer. No, you're not going to edit the next Pixar movie on it or explore the worlds of Elite Dangerous (but you could play the original!). A sense of perspective is what's needed. For the user who needs email, web access, and word processing or spreadsheet work, the price point is unbeatable. Even more advanced uses such as photo editing are certainly possible. If you're looking for a first computer for a young member of the family, then the price point (if they break it, it's not the end of the world), plus the possibilities afforded by the GPIO, make Raspberry Pi 4 well worth considering as their daily machine. **M**

▶ Watching YouTube videos proved a bit hit and miss

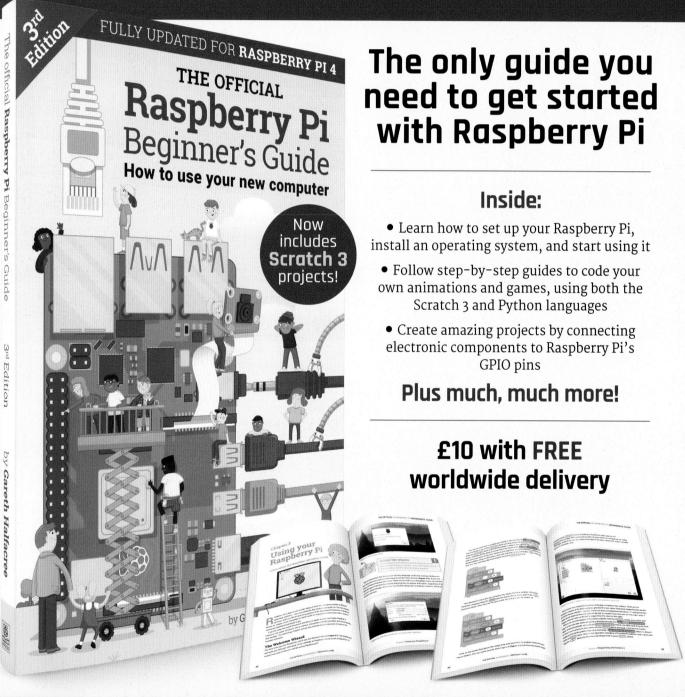

RASPBERRY PI 4K DIGITAL MEDIA HUB

Build the ultimate 4K home theatre PC using
a Raspberry Pi 4 and Kodi - By **Wes Archer**

We love Raspberry Pi for how it's helping a new generation of children learn to code, how it's resulted in an explosion of new makers of all ages, and how it's really easy to turn any TV into a smart TV.

While we always have a few Raspberry Pi computers at hand for making robots and cooking gadgets, or just simply coding a Scratch game, there's always at least one in the house powering a TV. With the release of the super-powered Raspberry Pi 4, it's time to fully upgrade our media centre to become a 4K-playing powerhouse.

We asked Wes Archer (**@raspberrycoulis**) to take us through setting one up. Grab a Raspberry Pi 4 and a micro-HDMI cable, and let's get started.

Get the right hardware

Only Raspberry Pi 4 can output at 4K, so it's important to remember this when deciding on which Raspberry Pi to choose

Raspberry Pi has been a perfect choice for a home media centre ever since it was released in 2012, due to it being inexpensive and supported by an active community. Now that 4K content is fast becoming the new standard for digital media, the demand for devices that support 4K streaming is growing, and fortunately Raspberry Pi 4 can handle this with ease! There are three versions of Raspberry Pi 4, differentiated by the amount of RAM they have: 1GB, 2GB, or 4GB. So, which one should you go for? In our tests, all versions worked just fine, so go with the one you can afford.

Remote controlled

It's possible to plug in an IR receiver and program Kodi so that you can use your TV remote to control your media centre, but the Flirc USB IR receiver is perfectly designed to do just that. Simply plug the USB receiver into a PC and follow a few quick steps to be up and running in no time.

Cases

Flirc Raspberry Pi 4 case

Made of aluminium and designed to be its own heatsink, the Flirc case for Raspberry Pi 4 is a perfect choice and looks great as part of any home media entertainment setup. This will look at home in any home entertainment system.

magpi.cc/NnDZiA

Official Raspberry Pi 4 case

The official Raspberry Pi 4 case is always a good choice. The snap-on top can easily be removed to access the Raspberry Pi board. If you're feeling adventurous, you can even hack the case to hold a small fan for extra cooling.

magpi.cc/frppYm

Aluminium Heatsink Case for Raspberry Pi 4

Another case made of aluminium, this is effectively a giant heatsink that helps keep your Raspberry Pi 4 cool when in use. It has a choice of three colours – black, gold, and gunmetal grey – so is a great option if you want something a little different.

magpi.cc/knNohY

Optional add-ons

▶ Maxtor 2TB external USB 3.0 HDD

4K content can be quite large and your storage
will run out quickly if you have a large collection.
Having an external hard drive connected directly
to your Raspberry Pi using the faster USB 3.0
connection will be extremely handy and avoids any
streaming lag.

magpi.cc/hyDQvY

◀ Fan SHIM

The extra power Raspberry Pi 4 brings means things
can get quite hot, especially when decoding 4K
media files, so having a fan can really help keep
things cool. Pimoroni's Fan SHIM is ideal due to
its size and noise (no loud buzzing here). There is a
Python script available, but it also 'just works' with
the power supplied by Raspberry Pi's GPIO pins.

magpi.cc/qiY6Wd

▶ Raspberry Pi TV HAT

If you are feeling adventurous, you can add a
Raspberry Pi TV HAT to your 4K media centre to
enable the DVR feature in Kodi to watch live TV.
You may want to connect your main aerial for
the best reception. This will add a perfect
finishing touch to your 4K media centre.

magpi.cc/imDdcw

▶ Rii i8+ Mini Wireless Keyboard

If your TV does not support HDMI–CEC, allowing
you to use your TV remote to control Kodi, then
this nifty wireless keyboard is extremely helpful.
Plug the USB dongle into your Raspberry Pi, turn
on the keyboard, and that's it. You now have a mini
keyboard and mouse to navigate with.

magpi.cc/AbrYu7

Get the right cables

Raspberry Pi 4 uses a micro-HDMI
cable instead of the standard HDMI.
In fact, it uses two as you can
output to two 4K displays at once.
We recommend buying the newer
micro-HDMI cables, but you can also
use micro-HDMI to HDMI adapters if
needed. If you do buy new cables,
be sure they are micro-HDMI to
HDMI, as most TVs will have the
standard-sized HDMI inputs.

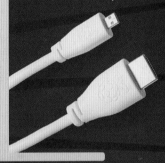

Install & set up LibreELEC

LibreELEC is a lightweight OS designed to run Kodi, the home media centre software we'll be using in this guide

Installation steps

01 Download the LibreELEC USB-SD Creator app

LibreELEC has a lovely app that makes it really simple to get up and running. Download the version for your OS (it supports Windows, Linux, and macOS) from **magpi.cc/epmapU** – or if you prefer, you can download the image for your chosen Raspberry Pi instead.

02 Download the LibreELEC image

Once downloaded, insert your microSD card into your computer and fire up the LibreELEC app you just downloaded. Select 'Raspberry Pi 4' from the version drop-down and then hit Download. Choose where to download the image file to and wait for the app to download the image.

03 Create your microSD card

Once your LibreELEC image has finished downloading, insert your microSD card into your computer, then select the drive from the drop-down menu. Lastly, hit Write and then wait for that to finish. Once done, you should have a working microSD card to use in your Raspberry Pi 4.

First-time setup steps

Spending some time **organising your libraries** first is definitely worth it

01 The first boot of LibreELEC

Now you have your microSD card ready, connect the micro-HDMI cable to your Raspberry Pi 4 and TV and then hit the power. The first time LibreELEC boots, it will do some housekeeping, like checking the file system and expanding to fill your microSD card before rebooting automatically for the next step.

02 Choose your language and get online

LibreELEC will launch a welcome wizard to help get you started. You'll need to pick your language, give your device a name (so it's easy to find on your network), connect to WiFi, and configure SSH and/or Samba file sharing. You should be able to control this using your TV's remote control without any setup if it supports HDMI-CEC.

> ❝ The first time LibreELEC boots, it will do some housekeeping, like checking the file system ❞

03 Welcome to Kodi on LibreELEC!

Once the wizard has been completed, you should be taken to the main home screen within Kodi. Congratulations! Right now, there's not much to do other than familiarise yourself with the menus, maybe make some settings changes (e.g. change the location to the correct one for you), and explore. We'll now show you how to add your media.

Add media libraries

01 Organise before you begin!

Before you add your media to your new Raspberry Pi 4K media centre, a little organisation is recommended. This way, when you add the libraries to Kodi, scrapers (more on that later) will be able to download all the extra artwork to make your entertainment system look the business.

02 Add your media libraries

A media centre is nothing without media, so we'll need to add ours before we can play them on our Raspberry Pi. Kodi makes this really simple, so navigate to Videos in the menu, select Files, and finally 'Add videos'. You'll now be able to use the Browse option to locate your media files, depending where they are stored.

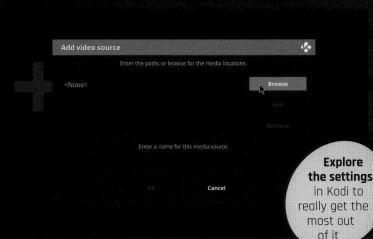

Explore the settings in Kodi to really get the most out of it

03 Give your libraries names

Organising media and scraping info

01 Using scrapers to add artwork

Scrapers are essentially scripts within Kodi that can search online databases to pull information about each media file you have, and download the corresponding artwork, such as movie poster, disc art, cases, and wallpapers (aka fan art). Assuming your libraries are organised as per our steps earlier, this should be straightforward.

02 Set the content type when adding media

When adding media, you should be asked what the directory contains (it defaults to 'None'). If you pick Movies from the options, this tells Kodi to use the appropriate information provider (i.e. The Movie Database) to scrape and download the resulting information and artwork when scanning the library.

03 Let the scraping begin!

Once you have set the content and chosen the information provider, hit OK and then you'll be asked if you want to refresh information for all items – hitting Yes will start the scraping, and you'll soon have detailed information about each movie, including any artwork too.

Advanced configuration

Now that the basics have been mastered in our Raspberry Pi 4K media centre, why not try something a little more advanced?

network-attached storage device (NAS for short) is a hard drive (or drives!), usually served by a lightweight operating system, that is attached to a network. The beauty is that you can share files across your network so that other network attached devices (such as our media centre) can access them.

Using your phone as the remote

Kodi has an official app available (iOS: **magpi.cc/LbeLJp**, Android: **magpi.cc/RBNqRy**) that allows you to control your media centre directly from your phone. Download the app to your phone, then hit 'Add host' to pair it with your Raspberry Pi. You can manually enter your Raspberry Pi's IP address, or hit 'Find Kodi' which should automatically find it for you. Once paired, you can use the app to navigate around Kodi like a regular remote.

Add network storage

01 Ensure SMB has been enabled in Kodi

Remember at the welcome wizard when you were asked about SSH and Samba? Well, Samba is not just a Brazilian dance, but a form of file sharing too! If you did not enable Samba (abbreviated to SMB), then you can do so in System > LibreELEC > Services.

02 Add your network shared media

This process is very similar as before. Go to Videos > Files > Add videos, then hit Browse. This time, select 'Windows network (SMB)' and then you should see your shared files appear. This assumes you have already configured SMB on your NAS; you may be prompted to enter a username and password if required.

03 Set the content for your libraries

Again, be sure to pick the content type for your network shared files. Kodi will add the files to the appropriate place in the navigation menus accordingly. Sharing files over the network can cause some buffering, depending on how fast and reliable your home network is, so just keep this in mind.

Advanced settings

▲ Update and clean your library automatically

If you regularly add and remove content to and from your libraries, you will want to ensure that Kodi updates the actual libraries too. Enable the 'Update library on startup' option to ensure that your new files are added automatically. It is also worth 'cleaning' your library too, in order to remove deleted files.

Set the region to your own

By default, Kodi favours the US audience, so if you are one of our US readers then this probably won't apply to you! However, the regional settings – such as the date, time, and temperature format – can be changed to match your own. Head to Settings > Interface > Regional and set it to your liking.

Change the look and feel of Kodi

There's nothing wrong with the default Kodi skin (Estuary), but there are a number of different ones to install and try out. Head over to Settings > Interface, then you can change the skin, or 'Get more' with a few clicks. Our personal favourite is Aeon Nox: SiLVO because there are so many customisations available.

▲ Enhance the information Kodi sees

Media files often have additional 'tags' that contain extra information about the file itself, which can be quite useful when you have a large collection. Going into Settings > Media > Videos and turning on 'Use video tags' enables this. Whilst you're there, ensure the three 'Extract…' options are on too, in order to enhance your experience.

App recommendations

Official Kodi Remote for iOS

Available on the App Store for iOS, the Official Kodi Remote is a sturdy choice for Apple users.

Kore, Official Remote for Android

Android users can also use the Official Kodi Remote app, called Kore, available in the Play Store.

Yatse

An alternative for Android users, Yatse has fantastic reviews and also has support for Plex and Emby servers.

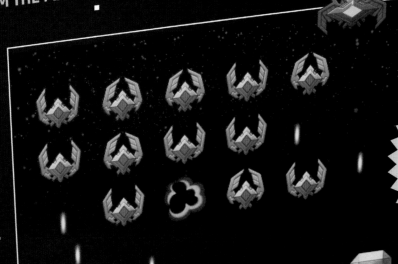

RETRO GAMING
WITH
RASPBERRY PI

Retro Gaming with Raspberry Pi shows you how to set up a Raspberry Pi to play classic games. Build your own portable console, full-size arcade cabinet, and pinball machine with our step-by-step guides. And learn how to program your own games, using Python and Pygame Zero.

- **Set up your Raspberry Pi for retro gaming**
- *Emulate classic computers and consoles*
- **Learn to program retro-style games**
- *Build a portable console, arcade cabinet, and pinball machine*

Project Showcases

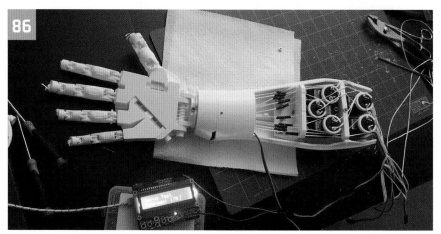

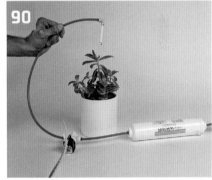

Project Showcases

As We Are

Ever wanted your face to be 14-foot high? Now you can with this sculpture covered in LEDs and powered by many, many Raspberry Pi boards. **Rob Zwetsloot** tries to get his head around it

MAKER

Matthew Mohr & Mac Pierce

Matthew is an artist who came up with the idea for As We Are, and Mac Pierce is the creative technologist that helped make it a reality.

magpi.cc/UjTkbW

Imagine walking through the doors of a convention centre and seeing someone's face staring down at you. Not just in statue form, or even on a big TV either – no, this is a 14 ft (4.3 m) high sculpture of a head, covered in LEDs that are hooked up to a load of Raspberry Pi boards to display a face. That must be somewhat disconcerting.

"The idea was concepted and artistically directed by Matthew Mohr," Mac Pierce, the production manager of the piece, tells us. "It was commissioned by the Greater Columbus Convention Center (**columbusconventions.com**) as the centrepiece of their new collection of local art."

It's certainly striking. One feature that is easy to miss is the Raspberry Pi-powered photo booth that is inside the head, allowing people to get their face scanned and the images 3D-mapped to the screens. It's not just showing off famous people, or even

Matthew the creator, but instead folks around you, or yourself. Mac opted to use Raspberry Pi boards in the booth after researching the standard solutions.

"Looking at network-attachable cameras, none of the versions we found would work for our application," says Mac. "Either they were scientific or industrial inspection cameras that required complex lensing and large mounting requirements – not to mention being prohibitively expensive – or they had impenetrable APIs and closed software that would limit their usefulness. The [Raspberry] Pi 3 and Camera Module combo ended up fitting nicely for our application, as the camera could be mounted separate from the main body of the [Raspberry] Pi, which allowed the overall size of the booth to have a lot less wasted space. The [Raspberry Pi boards] could also be easily attached to the network, and were open so that custom programs could be written on it to handle the photo processing and network handling."

At night, the head looks out on the road to intrigue passers-by

Quick **FACTS**

> There are 29 Raspberry Pi boards in the booth

> The process is similar to texture mapping on 3D models

> Creating maps in this way takes a lot of computing power

> The setup specifically uses a v2 Camera Module

> The head is built on castors which allow it to be rotated and moved easily

This 14 ft-tall display is made up of a series of many LEDs

Like any good art piece, this one has attracted visitors who would like to see and experience it

Mac explains, "The overall sculpture itself is built around an aluminium frame that gives the piece structure and [enables] mounting of the LEDs. The LEDs themselves were custom-manufactured by Sansi North

> **One feature that is easy to miss is the Raspberry Pi-powered photo booth inside the head**

America (**snadisplays.com**) to be able to form the curved layers of the head with only minimal faceting. The autonomous system of the scanning is done through a series of large processing servers, a half-rack of which exists in the top of the head, and another full rack that operates remotely."

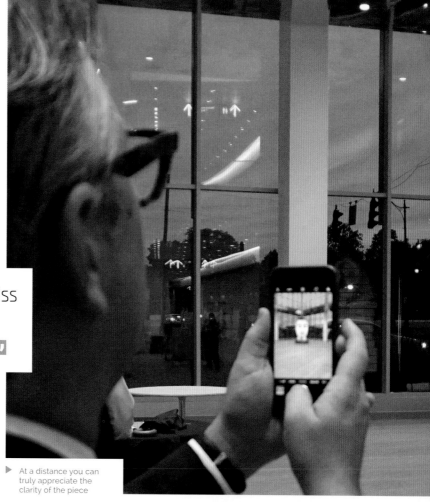

▶ At a distance you can truly appreciate the clarity of the piece

▼ Close up, you can see all the LED banks used to create the head image

▲ The interior frame is built of aluminium in slices like this

◀ These are individual LEDs, not just screens

▶ It takes a lot of Raspberry Pi boards to capture enough images for the project

The photo booth part operates via a touchscreen which guides the users on where to look and how to place their head for an optimal result. The Raspberry Pi boards then take the photos required by the servers.

See your reaction

The project has received a fair bit of attention, with one person calling it the 'Ultimate Selfie Machine', and it continues to operate in the Columbus Convention Center, where it remains open to the public. It even turns around to look out at the street at night.

As for future projects, projection mapping has taken Mac's interest: "I'm right now looking at a few different versions of projection-mapping software that runs on [Raspberry] Pi, which would make it perfect for permanently installed projection pieces. Also, there are a wealth of applications for [Raspberry] Pi in digital signage, of which I'm hoping to do some projects with soon." Ⓜ

Your face on the big screen

01 While you may be distracted by the head sculpture itself, make sure to walk around behind it to find a curtain. Walk through it to find where the magic happens.

02 Follow the on-screen instructions to take the perfect photo (or 30). These pictures are then sent to be processed by a server above you, and a server elsewhere.

03 Head back outside and you'll soon see your face looking back at you. Now's a perfect time to take a selfie of yourself with yourself.

Yuri 3 Mars rover

Airbus engineer John Chinner unveils the secrets behind his Pi-powered Yuri 3 Mars rover to a delighted **Rosie Hattersley**

MAKER

John Chinner

John is an engineer and STEM ambassador at Airbus UK. He worked on the Astro Pi project, testing the Raspberry Pi units sent into space.

magpi.cc/00maUK

▼ Yuri 3 sitting alongside Featherstone and Rocky Rover at the Pi Wars 2019 event
Credit: Harry Brenton

I n honour of the 50th anniversary of the Apollo moon landing, 2019's Pi Wars was space-themed. Visitors to the two-day event – held at Cambridge University at the end of March – were lucky enough to witness a number of competitors and demonstration space-themed robots in action.

Among the most impressive was the Yuri 3 mini Mars rover which was designed, lovingly crafted, and operated by Airbus engineer John Chinner. Fascinated by its accuracy, we got John to give us the inside scoop.

Airbus ambassador

John is on the STEM ambassador team at Airbus and has previously demonstrated its prototype ExoMars rover, Bridget (you can drool over images of this here: **magpi.cc/btQnEw**), including at the BBC Stargazing Live event in Leicester. Realising

> In Mission Mode, Yuri can be set to live-stream its journey across Mars's terrain while young scientists control it remotely

▲ John snuck Yuri 3 onto Airbus's Mars yard and gave it a test-run on a proper Mars test environment

the impressive robot's practical limitations in terms of taking it out and about to schools, John embarked on a smaller but highly faithful, easily transportable Mars rover. His robot-building experience began in his teens with a six-legged robot he took along to his technical engineering apprenticeship interview and had walk along the desk. Job deftly bagged, he's been building robots ever since.

> ❝ The part more challenging for home users is the 'gold thermal blanket' ❞

Yuri is a combination of an Actobotics chassis based on one created by Beatty Robotics (**beatty-robotics.com**), 3D-printed wheels, and six 12 V DC brushed gears. Six Hitec servo motors operate the steering, while the entire rover has an original Raspberry Pi B+ at its heart.

Yuri 3's Actobotics chassis is based on one used in rovers built by Beatty Robotics for a science museum

John designed, built, and 3D-printed the individually controllable wheels from scratch, while a Raspberry Pi B+ acts as Yuri 3's brain

Quick FACTS

> Prior to building Yuri 3, John was a Python novice

> Most of the development work took place on his mother-in-law's dining table!

> John was responsible for the shock, vibration, and EMC testing on the Astro Pi units

> Jim Henson's animatronic puppets were his original inspiration

> The gold blanket protects the rover in extreme conditions (as found on Mars)

Yuri 3 usually runs in 'tank steer' mode. Cannily, the positioning of four of its six wheels at the corners means Yuri 3's wheels can each be turned so it spins on the spot. It can also 'crab' to the side due to its individually steerable wheels.

The part more challenging for home users is the 'gold thermal blanket'. The blanket ensures that the rover can maintain working temperature in the extreme conditions found on Mars. "I was very fortunate to have a bespoke blanket made by the team who make them for satellites," says John. "They used it as a training exercise for the apprentices."

▲ Yuri 3 being tested on a beach to recreate challenging terrain navigation

Building a Mars rover

01 Yuri 3's chassis is based on a Beatty Robotics one, but the rover was designed and built by John. Some parts come from Actobotics, while the steering motors are standard RC-style servos. The gold blanket was bespoke-made for accuracy (but isn't required to function on Earth).

02 Yuri 3 has six separate servo motors, all controlled with Python. John operates the robot using a combination of Nintendo Wii Remote and Nunchuk controllers to manoeuvre the chassis and head respectively.

03 In 'mission mode' you get a rover-centric live-streamed view. Yuri follows commands entered by students in a classroom using the menu-driven Python script. Meanwhile, an audience watching in the school hall witnesses them attempt to avoid obstacles and seek a hidden alien.

▲ Yuri 3 explores the Martian landscape... only kidding – it's the local beach

▶ With its six individually steerable wheels and rugged chassis, Yuri 3 can handle rough terrain

John has made some bookmarks from the leftover thermal material which he gives away to schools to use as prizes.

Rover design

While designing Yuri 3, it probably helped that John was able to sneak peeks of Airbus's ExoMars prototypes being tested at the firm's Mars Yard. (He once snuck Yuri 3 onto the yard and gave it a test run, but that's supposed to be a secret!) Also, says John, "I get to see the actual flight rover in its interplanetary bio clean room".

His involvement with all things Raspberry Pi came about when he was part of the Astro Pi project, in which students send code to two Raspberry Pi devices on board the International Space Station. "I did the shock, vibration, and EMC testing on the actual Astro Pi units in Airbus, Portsmouth," John proudly tells us.

▼ John's daughter, and budding space-flight engineer inspects an earlier version of Yuri 3 – note the original red wheels

❝ What a fantastic opportunity for exciting outreach ❞

A very British rover

As part of the European Space Agency mission, ExoMars, Airbus is building and integrating the rover in Stevenage. "What a fantastic opportunity for exciting outreach," says John. "After all the fun with Tim Peake's Principia mission, why not make the next British astronaut a Mars rover? … It is exciting to be able to go and visit Stevenage and see the prototype rovers testing on the Mars Yard."

John also mentions that he'd love to see Yuri 3 put in an appearance at the Raspberry Pi store; in the meantime, drooling punters will have to build their own Mars rover from similar kit. Or, we'll just enjoy John's footage of Yuri 3 in action and perhaps ask very nicely if he'll bring Yuri along for a demonstration at an event or school near us.

John blogged the first year of his experience building Yuri 3 on his blog (**magpi.cc/oomaUK**). And you can follow the adventures of Yuri 3 over on Twitter (**@Yuri_3_Rover**). Ⓜ

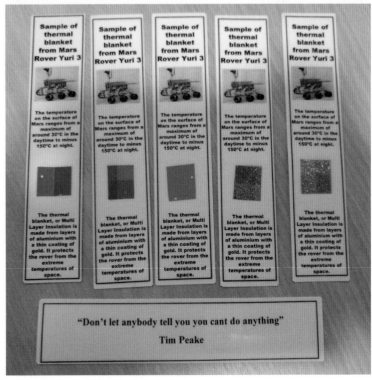

▲ The gold blanket is used to keep the Mars rover safe in extreme temperatures. Leftovers have been used to create bookmarks as rewards for students

SoFi

Developed by a team of researchers at MIT, this soft robotic fish swims alongside real ones in the ocean. **Phil King** dives in

MAKER

Robert Katzschmann

A PhD candidate at CSAIL, Robert is the lead author of the Robotic Fish for Underwater Exploration report. His previous projects include soft robotic arms and hands.

🔗 magpi.cc/kjsBkZ

I n the depths of the South Pacific, a strange new fish is exploring the Rainbow Reef.
Flexing its tail from side to side to propel itself serenely along, it captures the underwater scene using a camera – with a fish-eye lens! – mounted in its head, which also contains a Raspberry Pi 2 among other electronics.

This is SoFi (pronounced 'Sophie'), a soft-bodied robot created by researchers at MIT's Computer Science and Artificial Intelligence Laboratory (CSAIL) to study marine life up close, without disturbing it. That ingenious tail was inspired by the biological system used in tuna fins.

"The fish's motor pumps water into two balloon-like chambers in the tail," explains Robert Katzschmann, lead author of the project. "These work sort of like a pair of pistons in an engine: as one chamber expands, it bends and flexes to one side."

After working on SoFi and its predecessors for more than five years, Robert's team have perfected a naturalistic swimming action. "SoFi can turn, speed up, slow down, and move at different depths, including in strong currents," he reveals. "On average, SoFi swims at a speed of half a body length per second, though we plan to increase this further by improving its pump system and tweaking the design of its body and tail."

The robot has two fins on its side that adjust its pitch for diving up and down, while its overall buoyancy is controlled by an adjustable weight compartment and a chamber that can change its density by compressing and decompressing air.

Under pressure

"Among some of the challenges we encountered were the strong pressures that our fish had to withstand at deeper depths (down to 18 m) and the

SoFi swims by swishing its flexible silicone tail from side to side

A hydrophone receives ultrasonic signals sent by the diver's remote controller

SoFi's head features a camera linked to a Raspberry Pi hidden inside, protected by non-conductive oil

boundaries of our acoustic communication range for commanding SoFi remotely," says Robert. A diver uses a waterproof controller, containing another Raspberry Pi, to send commands to SoFi.

"Methods such as WiFi or Bluetooth don't work well underwater, so we chose to use sound instead," explains graduate student Joseph DelPreto. "The remote controller emits ultrasonic acoustic pulses

▲ SoFi's soft body and naturalistic propulsion system enable it to swim alongside other fish and marine animals without spooking them

> ## ❝ SoFi can turn, speed up, slow down, and move at different depths, including in strong currents ❞

that are too high-pitched for people to hear but that the robot can receive and decode to know how it should behave." The maximum control range is currently 20 m, but only reliable up to 10 m: "[It] could be higher but we wanted to minimise the disruption to other fish," says Robert.

Following fish

SoFi is also able to navigate autonomously to some degree using its on-board camera. "In the future we will show how SoFi can use its vision to follow other fish," reveals Robert. "By adding pre-recorded maps of the coral reefs onto the Raspberry Pi, we also plan to have the fish self-locate and navigate autonomously through the reefs."

Robert says the team hope to use SoFi to study deep-sea marine life that would be hard to capture otherwise. "The fish can not only gather video, but potentially also other sensor data, as well as taking water samples. We are for example curious to take water samples of the habitats, measure the temperature, and also record the sounds marine animals emit." M

Quick **FACTS**

> ➤ SoFi can operate at depths of up to 18 m

> ➤ A custom PCB is mounted on SoFi's Raspberry Pi

> ➤ An Mbed microcontroller handles the motors

> ➤ Ultrasonic echo is filtered by a custom algorithm

> ➤ Read the full paper at **magpi.cc/NEdPLE**

Swim like a fish

01 A gear motor pumps water alternately into two balloon-like chambers in the silicone elastomer tail, causing it to flex in one direction and then the other to mimic how a real fish moves.

02 Two side fins are angled to adjust SoFi's pitch to dive down and up. A buoyancy control unit uses compression to adjust the density of air inside it to determine SoFi's overall buoyancy.

03 Featuring a second Raspberry Pi connected to a HiFiBerry DAC+, the remote controller sends ultrasonic signals to SoFi. LEDs indicate the currently commanded state of the fish.

Teslonda

Jim Belosic and Michael Mathews have turned a Honda Accord into a hot rod gasser – powered by electricity and a Raspberry Pi. **David Crookes** gets up to speed

MAKER

Jim Belosic & Michael Mathews

Jim is the CEO and co-founder of the digital marketing platform ShortStack, and Michael is one of the lead software engineers at the company. Both love modding cars.

🔲 magpi.cc/uRHPfq

Many people remember their first car, but very few will keep hold of it for decades. Then again, not everyone is like Jim Belosic, a vehicle modder who saw great potential in his 1981 Honda Accord. Rather than sell it, he decided to bring it into the 21st century. And that meant fitting it with an electric motor from a salvaged Tesla Model S P85.

"I've been wrenching on something since I can remember," Jim says. "So turning the Accord into an electric car seemed like a good way of keeping it around for the nostalgia. I also figured that if I want to be able to modify cars in the future, I'd better learn everything I can about these kinds of vehicles now."

The work involved replacing the steering and suspension system and moving to a straight-axle front end to accommodate the battery pack. Jim also added some drag-race tyres. "It gave the car a ton of character," he says. But what makes the car rather special for us is Jim's integration of a Raspberry

Pi 3. This was carried out by his car-modding and software-developing friend, Michael Mathews.

Motor monitoring

As well as powering the electronic dash, their Raspberry Pi allows for feedback and configuration of the motor hardware. It can constantly monitor the temperature levels of the batteries and motor to ensure the car is not being overexerted, and it can be used for both traction control and to change the voltage and amperage levels to the motor.

"For this particular project, I wanted to dive straight into the HTML5 Canvas element [which can draw graphics on the fly via JavaScript] because I figured if I could control how and when something was drawn to the screen, I could get it to run pretty decently on [Raspberry] Pi," Michael explains. "I could get a prototype up and running fairly quickly using web tech."

One main goal was to make it semi-portable so that it could be stuffed into another project without much rework. "I also wanted to allow any device to connect to it via WiFi through a web server, and it needed the ability to monitor, control, and log data on the back end through a web app," Michael continues.

Canvas frames

To achieve all of this, Michael grabbed a pencil and paper to draw his desired UI and design flow. He opted to use the Chromium browser in its kiosk mode running an accelerated Canvas, and he wanted the back end to listen to data from the motor's controller for incoming messages using a Node.js server via a WebSocket. At first, he had problems with the visual performance since he was getting below 20 frames per second and heavy spikes of lag.

He solved this by using a 'frame' to only get the back end to send the most updated data rather than every single message. He also minimised browser reflows, and enabled Canvas acceleration

▲ The Teslonda is a cross between a 1981 Honda Accord, a Tesla Model S P85, and a 1960s gasser-style dragster. And it's extremely quick

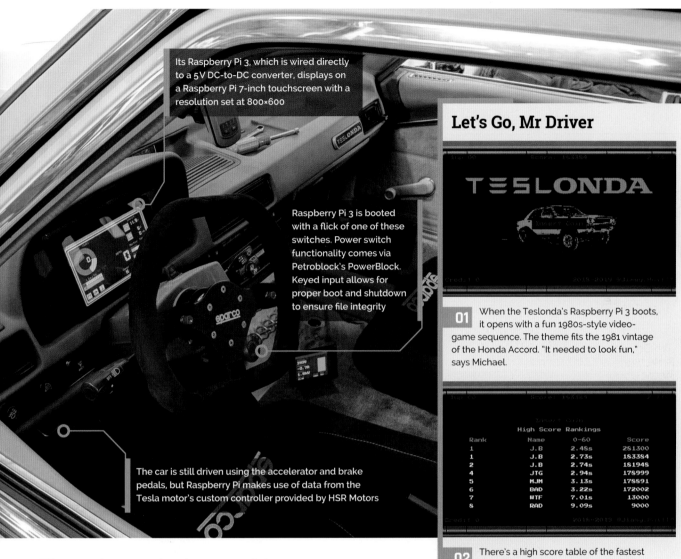

Its Raspberry Pi 3, which is wired directly to a 5V DC-to-DC converter, displays on a Raspberry Pi 7-inch touchscreen with a resolution set at 800×600

Raspberry Pi 3 is booted with a flick of one of these switches. Power switch functionality comes via Petroblock's PowerBlock. Keyed input allows for proper boot and shutdown to ensure file integrity

The car is still driven using the accelerator and brake pedals, but Raspberry Pi makes use of data from the Tesla motor's custom controller provided by HSR Motors

Let's Go, Mr Driver

01 When the Teslonda's Raspberry Pi 3 boots, it opens with a fun 1980s-style video-game sequence. The theme fits the 1981 vintage of the Honda Accord. "It needed to look fun," says Michael.

02 There's a high score table of the fastest 0 to 60. "I did some research on what early 1980s digital dashboards looked like and found my inspiration – a Mitsubishi Cordia 1982 digital dash."

03 The dash shows the speed, gear, voltage, amp, temperature, power, and more. "I wanted the driving experience to feel like you're at the arcade. There's also a 'Continue?' countdown when the Tesla motor is turned off."

❝ Raspberry Pi 3 boots with a fantastic display that's reminiscent of a 1980s arcade game ❞

by turning on every relevant flag on Chromium. By making sure Canvas would only draw when something updated, and erased only what was dirty, he could maintain 45 to 60 fps on the dash.

The result is amazing. Turn the car on, flick a switch to activate the dash, and its Raspberry Pi 3 boots with a fantastic display that's reminiscent of a 1980s arcade game. It shows the speed, battery voltage, charge, and temperature, among other attributes. "I'll soon be hooking up a GPS to assist with logging, acceleration, and G-force readings as well," Michael reveals. "And that's the only problem: I now want to do so much more." ❚

Quick **FACTS**

> The car is operated by toggle switches

> Jim wants to let it rip on the drag strip

> The software will be licensed for others to use

> A mobile web app can log and track functionality

> The car can go from 0–60 in 2.48 seconds

One part workout, one part game. Defeat your adversary as you punch

Pi Fighter

Gamifying boxing with a special punchbag that allows you to fight...
Luke Skywalker? **Rob Zwetsloot** starts a training montage to check it out

MAKER

Richard Kirby

A test manager for the London Underground by day, and Raspberry Pi tinkerer by night. Richard also runs the London Raspberry Pint meet-ups.

■ magpi.cc/rvxvXT

Did you know that the original version of **Street Fighter had a variant where you could punch the buttons to get Ryu to attack?** The harder you smacked the kick button, the more damage it would do. These apparently wore out very quickly, which is why watching Street Fighter tournaments these days is akin to watching someone playing the piano. Albeit with six buttons and a joystick.

What if you could bring this back? And combine it with other arcade classics and staples? Meet Richard Kirby's Pi Fighter.

A new challenger!

"Pi Fighter is essentially a real-world old-school fighting video game," Richard tells us. "The player chooses an opponent and challenges them to a sparring match. Each player has a certain number of health points that decrement each time the other player lands an attack. Instead of clicking a joystick or mouse button, the player hits a heavy bag. The strength of the hit is measured by an accelerometer. [A Raspberry] Pi translates the acceleration of the heavy bag (measured in G) into

> ❚❚ The player hits a heavy bag. The strength of the hit is measured by an accelerometer ❚❚

▲ 3… 2… 1… Fight!

Keep track of your health with the display. Don't punch it, though

Quick **FACTS**

> Pi Fighter only uses three electronic components

> Richard warns against the tricky Jedi fighters

> Raspberry Pint can be found at CodeNode in London

> The lack of moving parts made a Raspberry Pi perfect for the job

> A heavy bag is recommended over a wall bag

Use a heavy bag to get a good workout and a good idea of your punch strength, *Rocky IV* style

Go for broke!

01 Step up to the bag. Your opponent has been selected and you need to defeat them before they can drain your HP.

02 Hit the bag hard and fast. The stronger your punch, the more damage your opponent receives. They do fight back, although at least they don't actually hit you.

03 Once you've defeated your foe, your HP recovers a bit. No time to rest, though: your next opponent awaits. Do you have what it takes to fight Darth Vader?

▲ The setup is remarkably simple. It's all in the code

the number of health points to decrement from the opponent. [Raspberry] Pi runs your opponent, which attacks you – you don't actually get hit, but your health points decrement each time they attack."

It's a remarkably simple idea, and it started off as just an app that used a smartphone's accelerometer. Translating that to a Raspberry Pi is just a case of adding an accelerometer of its own.

"I realised it could be used to measure the overall strength of a punch, but it was hard to know how that would translate into an actual punch, hence the idea to use a heavy bag," Richard explains. "This appealed to me as I studied karate and always enjoyed hitting a heavy bag. It is always difficult to gauge your own strength, so I thought it would be useful to actually measure the force. The project ended up

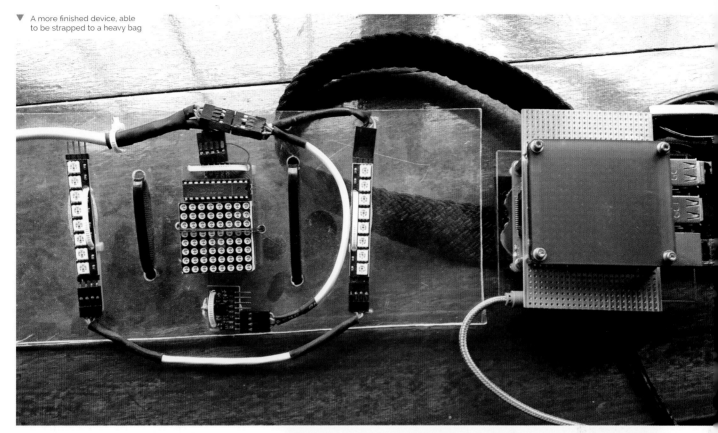

▼ A more finished device, able to be strapped to a heavy bag

❝ I thought it would be useful to actually measure the force ❞

consuming a good amount of time, as you would expect when you are learning."

Finish them?

While Pi Fighter is already used at events, Richard says, "It needs a bit of tuning and coding to get everything right [...]. It could be a never-ending project for me. You can always fix things and make the software more robust, the user interface more usable, etc. It isn't mass rollout ready, but I have never had it fail at a key moment such as presenting at a Raspberry Jam or Raspberry Pint. It (mostly) gets better every time I put some effort into it."

If you find yourself at Raspberry Pint in London (**@raspberrypint** on Twitter), make sure to do a bit of a warm-up first – you might find yourself head-to-head in a boxing match with a Jedi. Here's hoping they don't know Teräs Käsi. ▨

▲ Not many makers get to enjoy punching their own project

The Squirrel Cafe

Predict the weather with... squirrels and nuts!?
Nicola King lifts the lid on an ingenious project

MAKER

Carsten Dannat aka 'The Squirrel Gastronomer'

A software engineer from Ahrensburg in Germany, Carsten opened the Squirrel Cafe in 2007, later adding the Raspberry Pi IoT project to monitor it.

thesquirrelcafe.com

Back in 2012, Carsten Dannat was at a science summit in London, during which a lecture inspired him to come up with a way of finding correlations between nature and climate. "Some people say it's possible to predict changes in weather by looking at the way certain animals behave," he tells us. "Perhaps you can predict how cold it'll be next winter by analysing the eating habits of animals? Do animals eat more to get additional fat and excess weight to be prepared for the upcoming winter?"

An interesting idea, and one that Germany-based Carsten was determined to investigate further. "On returning home, I got the sudden inspiration to measure the nut consumption of squirrels at our squirrel feeder", he says. Four years later and his first prototype of the 'The Squirrel Cafe' was built, incorporating a first-generation Raspberry Pi.

A tough nut to crack

A switch in the feeder's lid is triggered every time a squirrel opens it. To give visual feedback on how often the lid has been opened, a seven-segment LED display shows the number of openings per meal break. A USB webcam is also used to capture images of the squirrels, which are tweeted automatically, along with stats on the nuts eaten and time taken. Unsurprisingly perhaps, Carsten says that the squirrels are "focused on nuts and are not showing interest at all in the electronics!"

So, how do you know how many nuts have actually been eaten by the squirrels? Carsten explains that "the number of nuts eaten per visit is calculated by counting lid openings. This part of the source code had been reworked a couple of times to get adjusted to the squirrel's behaviour while grabbing a nut out of the feeder. Not always has a nut been taken out of the feeder, even if

Quick **FACTS**

> It correctly predicted the cold winter of 2017-18

> Carsten plans to add a scale to weigh the nuts...

> ...for more accurate measuring of nut consumption

> Raccoons have broken into the feeder

> A video 'security camera' now monitors all visitors

A mechanical switch is pressed whenever the lid is opened by a squirrel

A Raspberry Pi is connected to the switch, LED display, and a USB webcam to take photos

The nuts are visible behind a glass panel

the lid has been opened." Carsten makes an assumption that if the lid hasn't been opened for at least 90 seconds, the squirrel went away. "I'm planning to improve the current design by implementing a scale to weigh the nuts themselves to get a more accurate measurement of nut consumption," he says.

Just nuts about the weather!

The big question of course is what does this all tell us about the weather? Well, this is a complicated area too, as Carsten illustrates: "There are a lot of factors to consider if you want to find a correlation between eating habits and the prediction of the upcoming winter weather. One of them is that I cannot differentiate between individual squirrels currently [in order to calculate overall nut consumption per squirrel]."

> **Some people say it's possible to predict changes in weather by looking at the way certain animals behave**

He suggests that one way around this might be to weigh the individual squirrels in order to know exactly who is visiting the Cafe, with what he intriguingly calls "individual squirrel recognition" – a planned improvement for a future incarnation of The Squirrel Cafe.

Fine-tuning of the system aside, Carsten's forecast for the winter of 2017/18 was spot-on when he predicted, via Twitter, a very cold winter compared to the previous year. He was proven right, as Germany experienced its coldest winter since 2012. Go squirrels!

▲ The results of a raccoon's rampage

▼ A squirrel enjoying a tasty treat at the Squirrel Cafe

Secret squirrel

01 When a squirrel opens the lid, a mechanical switch is triggered. This replaced a magnetic reed switch, which Carsten says wasn't totally reliable.

02 The feeder is filled with peanuts. Since these are split into halves, it's assumed that each lid opening results in half a nut being consumed by the squirrel.

The Squirrel Cafe @TheSquirrelCafe · Jun 22
#Squirrel chowed down on 3.5 nuts for 3.43 min at 17:38:08 CEST. An #IoT project to predict how cold it'll be next winter. #ThingSpeak

03 After each meal visit, the Tweepy Python library is used to automatically tweet the details and a photo taken by a connected USB webcam.

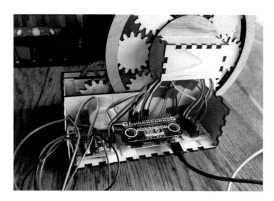

◀ Assembling the laser-cut hub and planetary gears, which are rotated by a stepper motor

◀ A Raspberry Pi Zero runs the code, a Speaker pHAT plays chime sounds, while a Trinket M0 controls the NeoPixels

Tide Clock
Weather Thing

This beautifully crafted device provides a detailed weather forecast and low tide times. **Phil King** is on cloud nine

MAKER

Fin Hopkins

A software engineer on the City of Boston's Digital Team, Fin is into board games, web development, and social justice.

magpi.cc/DtBrR7

Seeking to make a gift for the in-laws looking after the kids in Maine over the summer, Fin Hopkins decided to build a 'Tide Clock Weather Thing' to help predict the weather and tides for days out at the beach or kayaking.

"I remember [the reaction from the recipients] was something like 'wow, that's beautiful! What is it?' I had to point out what all the lights and dials were, since there aren't any markings on the case," recalls Fin.

> ❝ I had to point out what all the lights and dials were, since there aren't any markings on the case ❞

How it works

The large wheel on the device shows the current weather conditions on top; as they change, it rotates planetary gears to bring a new icon to the top. In the middle of the wheel, a finger points to the current temperature, with the forecasted daily range lit by coloured LEDs. Five more LEDs below light up blue for impending rain, filling up to show when it's 60, 45, 30, 15, or 5 minutes away.

24 LEDs at the bottom of the device represent each hour of the day, lighting up in different colours for forecast weather conditions – including

blue for rain, yellow for sunny, dim white for cloudy, and green for windy. Just above this strip, a moving bar with two pointers shows when the two daily low tides will occur.

A chime is also sounded for low and high tides using a Speaker pHAT connected to a Raspberry Pi Zero, which runs the Python software and controls the NeoPixel LEDs via a Trinket M0 microcontroller. All the weather data is sourced from the Dark Sky API, while tide data comes from the NOAA's Tides and Currents site.

Time to make

From the first gear prototypes to a working version of the device took Fin about three weeks, working nights and weekends. "We were going to visit my in-laws for the 4 July holiday, so it was a sprint to a tight deadline at the end. I got all the hardware and wiring done, and then ended up finishing the coding while I was up there."

The design adapted and fell into place as Fin prototyped, starting with the laser-cut planetary gears – for the weather symbols – which rotate around a central hub. "Once I saw how neat the large version [of the wheel] looked, it was 'OK, what else can I put with this?'"

A lot of the changes made concerned reducing the scope. "I wanted to show high and low tides over two days, with each indicator separately

Quick **FACTS**

> It's made from medium cherry and maple plywood

> The large wheel rotates 360° at the top of every hour

> A NeoPixel ring provides the LED arcs

> The laser-cut parts were designed in Inkscape

> Fin is working on an edge-lit acrylic sign

The weather icons (designed by Austin Condiff, CC BY) are printed on planetary gears that rotate as they move

The finger points to the current temperature, while lit LEDs show the daily range in 5°F increments

A micro-servo and gears move two pointers along to indicate times for the day's low tides

controlled. Once I got into 'OK, now how would I build that?' it became one day, then not separately controlled, then just low tides," explains Fin. "Even after I simplified my goal for the tide indicator, it took a number of prototypes to get right."

Making their debut Raspberry Pi – and electronics – project was an educational process for Fin. "I learned how to solder, how to crimp wire, and I finally got my head around the difference between voltage and current. I learned about stepper motors, GPIO pins, I²C communication, PWM, and a ton of other topics." M

◀ A top-down view of the project, showing how all the laser-cut and electronic parts fit together

A Whisper of Moths

An art installation filled with hand-crafted moths that move. **Rob Zwetsloot** looks around it

MAKER

Macclesfield Community ArtSpace

A charity based in Macclesfield and run entirely by volunteers, it's a free workshop full of multitalented members able to create A Whisper of Moths.

magpi.cc/WNeWNR

We always love to see art installations that make use of the Raspberry Pi, and A Whisper of Moths is a perfect example. This particular project involved several people from the Print Mill, which is part of Macclesfield Community ArtSpace, along with technical and labour support from IDST! (If Destroyed Still True!), another ArtSpace group.

"The moths were made by local schoolchildren using recycled A4 plastic wallets that were drawn on and/or filled with glitter, ironed between baking parchment and cut to shape," Nick Young from the team tells us. "Some moths were also 3D-printed."

The movement of the moths was simulated by projecting randomly positioned and sized circles onto the moths, which were hanging by fishing line from fine garden netting suspended from the church balcony.

▶ The moths were made using recycled, or by recycling, materials

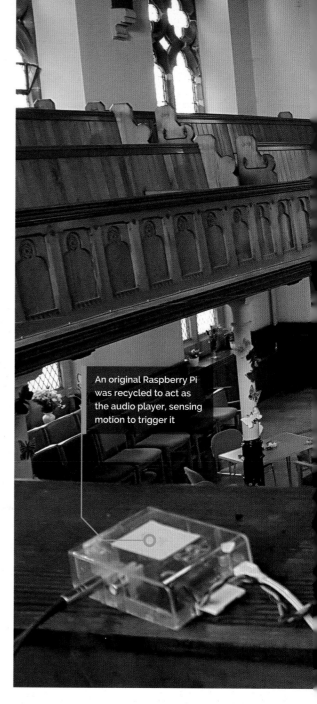

An original Raspberry Pi was recycled to act as the audio player, sensing motion to trigger it

The installation was interactive, with the sound of whispers being triggered as people walked through, along with the projected circles of light to simulate movement to some degree. How did such an idea come about, though?

Lunar new year

"The Town Council approached Macclesfield Community ArtSpace wishing to celebrate in some form, Chinese New Year," Nick explains. "Macclesfield's link with China is through silk and therefore the suggestion was made to create an installation that incorporated silk moths and engaged the local schoolchildren in the making process. Macclesfield is also striving to draw attention to the problems of single-use plastic and so we chose to use plastic and recycle it."

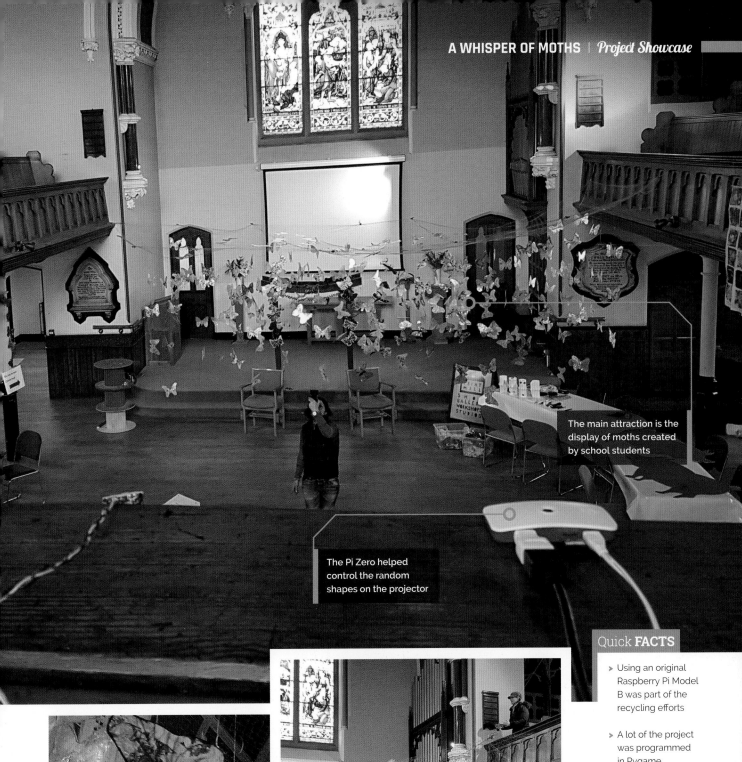

The main attraction is the display of moths created by school students

The Pi Zero helped control the random shapes on the projector

Quick **FACTS**

> Using an original Raspberry Pi Model B was part of the recycling efforts

> A lot of the project was programmed in Pygame

> The sounds have been reused for Halloween and Christmas

> We're currently in the Year of the Pig, hence the papier mâché pig

> Macclesfield was once the world's largest producers of finished silk

▲ Even without the nice stained-glass light, the butterflies themselves are very colourful

▶ The project setup was quite involved – a lot of the church was used!

Whispering to moths

01 As visitors walk into the church, the PIR sensor picks them up and begins the A Whisper of Moths sequence.

02 The original Pi Model B begins to play a random sample of whispering sound effects, which is how the project gets its name.

03 As the audio plays, random circles of light are projected onto the moths hanging above the visitors, which – along with the whispering – gives an illusion of movement.

▲ A dragon dance led the visitors to the church, where they would experience A Whisper of Moths

After following a Chinese dragon puppet and papier mâché pig down the high street in a parade during the Chinese New Year festivities, people walking into the church would trigger the display. "We had the added and unforeseen bonus of coloured light filtering through the moths as the sun shone through the huge stained glass windows," Nick recalls.

Raspberry Pi connection

"We discussed motorising some or all moths, even just with vibration motors, but discarded this as inappropriate for a sustainable art-piece," Nick says. "We decided to have a moving pattern of light [shone] around or onto the static exhibit to create the illusion of movement. We created a prototype using metres of LEDs, but these were not bright enough and on surveying the venue, an active local church, we realised that we could use a projector to beam the light onto the art.

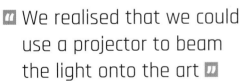

> ❝ We realised that we could use a projector to beam the light onto the art ❞

"We used the first Raspberry Pi for the sound because the simple movement sensor could trigger and play many sound files, reusing an existing setup, and a second Raspberry Pi Zero also programmed with Pygame to display a series of random white circles against a black background."

The final setup was pretty simple, making use of an old Raspberry Pi Model B, a Raspberry Pi Zero W, and PIR motion sensor, plus the speakers and projection equipment.

Currently, the exhibition has been taken down. However, the team are looking to set it up in a local museum soon. ◢

▲ Each butterfly was painstakingly added to the net

◀ The stained glass windows added some unexpected colour to the moths

Underwater Drone

Ievgenii Tkachenko not only took the plunge with a Raspberry Pi but with his first Raspberry Pi project too, as **David Crookes** explains

MAKER

Ievgenii Tkachenko

Ievgenii is a senior Android developer from Kyiv, Ukraine. As well as writing code, he loves engineering challenges and creating something interesting and helpful.

magpi.cc/GRSQHi

Never let it be said that some makers won't jump in at the deep end for their ambitious experiments with the Raspberry Pi. When Ievgenii Tkachenko fancied a challenge, he sought to go where few had gone before by creating an underwater drone, successfully producing a working prototype that he's now hard at work refining.

Inspired by watching inventors on the Discovery Channel, Ievgenii has learned much from his endeavour. "For me it was a significant engineering challenge," he says, and while he has ended up submerging himself within a process of trial-and-error, the results so far have been impressive.

Raspberry Pi dive

The project began with a loose plan in Ievgenii's head. "I knew what I should have in the project as a minimum: motions, lights, camera, and a gyroscope inside the device and smartphone control outside," he explains. "Pretty simple, but I didn't have a clue what equipment I would be able to use for the drone and I was limited by finances."

Bearing that in mind, one of his first moves was to choose a Raspberry Pi 3B, which he says was perfect for controlling the motors, diodes, and gyroscope while sending video streams from a camera and receiving commands from a control device.

► The LEDs are attached to radiators to prevent overheating, and a pulse driver is used for flashlight control

▲ The project's Raspberry Pi 3 sits in the housing and connects to a LiPo battery that also powers the LEDs and motors

"I was really surprised that this small board has a fully functional UNIX-based OS and that software like the Node.js server can be easily installed," he tells us. "It has control input and output pins and there are a lot of libraries. With an Ethernet port and WiFi and a camera, it just felt plug-and-play. I couldn't find a better solution."

Working with a friend, Ievgenii sought to create suitable housing for the components, which included a twin twisted-pair wire suitable for transferring data underwater, an electric motor, an electronic speed control, an LED together with a pulse driver, and a battery. Four motors were attached to the outside of the housing and efforts were made to ensure it was waterproof. Tests in a bath and out on a lake were conducted.

Streaming video

With a WiFi router on the shore connected to the Raspberry Pi via RJ45 connectors and an Ethernet cable, Ievgenii developed an Android application to connect to the Raspberry Pi by address and port ("as an Android developer, I'm used to working with the platform"). This also allowed movement to be controlled via the touchscreen, although he says a gamepad for Android can also be used. When it's up and running, Raspberry Pi streams a video

Wires leave the housing through cable glands and sit in the water, but they have been sealed

The camera was placed in this transparent waterproof case attached to the front of the waterproof housing

The drone uses four brushless motors – two positioned horizontally and two vertically. Ievgenii would like to add two more verticals for better control

Quick **FACTS**

> The prototype worked well at a depth of two metres

> It can be controlled via an Android app

> A lot of testing was required

> It operated best in a calm lake

> Sand and algae can clog the motors

from the camera to the app – "live video streaming is not simple and I spent a lot of time on the solution" – but the wired connection means the drone can only currently travel as far as the cable length allows.

In that sense, it's not perfect. "It's also hard to handle the drone and it needs to be enhanced with an additional controls board and a few more electromotors for smooth movement," Ievgenii admits. But as well as wanting to base the project on fast and reliable C++ code and make use of a USB 4K camera, he can see the future potential and he feels it will swim rather than sink.

"Similar drones are used for boat inspections and they can also be used by rescue squads or for scientific purposes," he points out. "They can be used to discover a vast marine world without training and risks too. In fact, now that I understand Raspberry Pi, I know I can create almost anything, from a radio electronic toy car to a smart home." 🅜

▲ The Underwater Drone being tested out in the shallows

Chord Assist

With a built-in AI voice assistant, this accessible smart guitar can help almost anyone learn to play. **Phil King** starts strumming

MAKER

Joe Birch

Joe is an Android engineer working remotely at Buffer, currently from Brighton in the UK. He is also a Google Developer Expert in Android, Google Pay, and Flutter.

chordassist.com

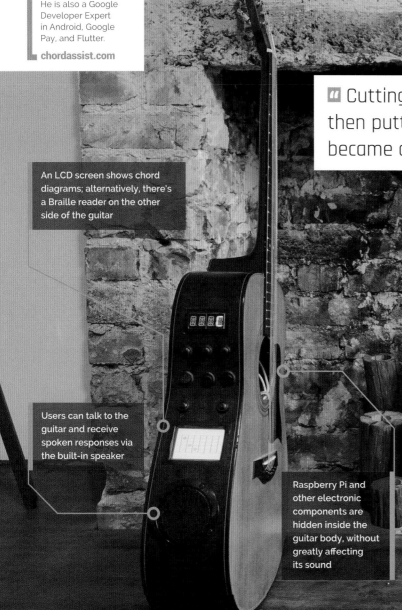

An LCD screen shows chord diagrams; alternatively, there's a Braille reader on the other side of the guitar

Users can talk to the guitar and receive spoken responses via the built-in speaker

Raspberry Pi and other electronic components are hidden inside the guitar body, without greatly affecting its sound

Learning to play the guitar can prove difficult for people with sight or hearing loss. Joe Birch's accessible Chord Assist guitar is intended to make the process a whole lot easier for deaf, blind, and mute people.

In Joe's family runs a condition known as retinitis pigmentosa, which causes tunnel vision over time; it has affected his mother, who is now registered partially sighted. "Being closer

> **"** Cutting the holes in the guitar and then putting all of the parts inside became quite a tedious task **"**

to people who have this condition opened up my awareness of how it can effect peoples lives," explains Joe. "Currently, music is something that is not so accessible to everyone, so I started to think of ways in which it could become accessible – which is where the idea for Chord Assist came from."

As well as an LCD screen, four-digit display, and buttons to show and select chords, Joe's modified acoustic guitar features a Braille reader based on his earlier BrailleBox news-reader project (**magpi.cc/LpjrGF**). There's even a vibrating progress indicator next to the reader, to indicate when a request is taking place.

In addition, the user can make a spoken request for a chord using the built-in mic and Google-based voice assistant, and hear a response via the speaker. Also capable of playing a requested note for tuning purposes, this system makes use of Joe's Chord Assist app (**magpi.cc/xUAQNo**) on the Actions on Google platform, which can also be used as a standalone learning aid on smartphones and other devices.

Hidden components

The whole project took Joe around six months to complete. "The most difficult part was definitely the last stages of the build process –

The Braille reader on the other side of the guitar enables blind users to read instructions on how to play a chord

cutting the holes in the guitar and then putting all of the parts inside became quite a tedious task," he recalls. "I originally had everything on prototyping boards, but components kept coming unplugged. Because of this I decided to solder everything properly on PCB boards, which improved everything here."

All the electronic components, including a Raspberry Pi, are secreted inside the body of the guitar. "There's actually a lot of stuff in there, but none of it is that heavy other than the portable battery pack." It doesn't seem to have an adverse effect on the sound quality either: "Initially I thought it was going to be a big problem, but when comparing it against my other guitar, there isn't really much difference."

Turn it up

While Joe is happy with the project, he's planning a few tweaks. "I'd like to add an amplifier and volume control to improve the functionality of the speaker. Another thing I'd like to add is real guitar tuning functionality – so something in the software that will analyse the note played and let the user know if the string needs to be tuned higher or lower. This isn't yet possible with the Google Assistant, but would be a useful feature for users who might not have experience with tuning a guitar. I could also make use of the screen to instruct the user here and cater for the different accessibility needs."

If you'd like to have a go at creating a similar project, Joe has open-sourced most of what is needed to build the guitar, apart from the program that runs on Raspberry Pi which drives the user interface and GPIO for the components. "The wiring diagram and conversational bot are open-source, so someone could build it if they wanted to."

Following feedback from users, Joe plans to iterate on the design to create more smart guitars. Since this project has struck a chord with us, that's music to our ears! ⓜ

Raspberry Pi and other components, including an eight-channel relay for the solenoids in the Braille reader, are stuffed inside the body of the guitar

Extensive controls and features include selection buttons, a mic and speaker, along with a segment display and LCD screen

Quick **FACTS**

> Chord Assist has been nominated for a Webby award

> The screen, Braille reader, and speaker can each be turned on/off individually

> There's a button to quickly repeat the last returned voice response

> A four-digit segment display shows available chords...

> Buttons below it can be used to select a chord, instead of speaking

MAKER

Johanna Tano

Self-taught programmer Johanna's visual installations have appeared at music festivals, in a forest in Sweden, and at fashion shows. Each is controlled by Raspberry Pi.

▮ johannatano.com

Warning!
High voltage

CRT TVs contain high voltages, so be very careful if using them

TV Wall

Fans of analogue TVs and upcycling old tech will be wowed by this visually arresting Raspberry Pi-controlled display, reckons **Rosie Hattersley**

Where would the Wizard of Oz have been without the visual artifice that kept his myth alive? Being confronted with Johanna Tano's wall of TV screens surely has a similar disembodied effect to the sight that greeted Dorothy and pals when they entered the Emerald City. Harnessing the power of Raspberry Pi computers, Johanna has been able to sync up and independently control up to 30 analogue TV displays at once.

In her aptly named TV Wall, Johanna demonstrates the possibilities of using old tech and new in a highly engaging fashion. The TV Wall had its debut at Stockholm's Fashion Week in 2017, where the likes of singer Kelis partied to a backdrop of multi-screen live video mash-ups.

The sun always shines on TV

In its most recent iteration, the TV Wall is central to Swedish singer Zacharias Zachrisson's music video for *Shadow*. He says: "Together with the video director, Albin Eidhagen, we created custom-made

videos and animations for the TV screens, which was live-mixed when we filmed it." The result is a decidedly eighties video, not unlike something by A-Ha.

Appropriately enough, the video came about when the singer saw the TV Wall displaying live code-generated visuals on Instagram and got in touch with Johanna. She had only a week to assemble the build for the music video shoot and says things "get complicated, fast" as the installation is scaled up and more TVs – and more IP addresses – are added.

Those inspired by her project can follow in her footsteps by breaking it down. "Figure out how to get a video signal from Raspberry Pi to a TV, then how to stream video from a computer to a Raspberry Pi," she advises. "Work out how to tell multiple Raspberry Pi boards to display different parts of the same video. By solving each step, you end up having a quite advanced system without even realising it."

> ▮▮ Work out how to tell multiple Raspberry Pi boards to display different parts of the same video ▮▮

▶ Johanna sourced the analogue TVs from a range of online sources, paying about 100 euros for each

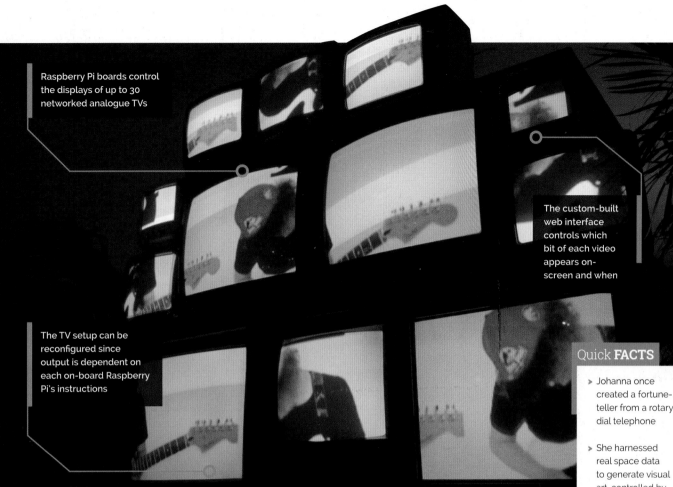

Raspberry Pi boards control the displays of up to 30 networked analogue TVs

The custom-built web interface controls which bit of each video appears on-screen and when

The TV setup can be reconfigured since output is dependent on each on-board Raspberry Pi's instructions

Quick **FACTS**

> Johanna once created a fortune-teller from a rotary-dial telephone

> She harnessed real space data to generate visual art, controlled by an EEG

> She was part of the Berlin-based tech-art collective Lacuna Lab

> Her first Raspberry Pi project lit up Sweden with 45 LED light trees users controlled remotely

> She views Raspberry Pi as the ideal way to control hardware used in conceptual art projects

▶ Some impressive visual effects can be achieved by sending a different video feed to each TV

◄ Automatic mapping is controlled through the web interface, meaning she can easily modify each Raspberry Pi's settings, including which part of each video to use

▼ A TCP Syphon Server application connects to the TV Wall network, enabling the video or visual output generated by the VJ software to be displayed in real-time on the wall

Raspberry Pi on TV

01 Cables, adapters, network switches, and routers, plus the customisable TV stands, cost Johanna around 500 euros.

02 Johanna tried to make the monster stand a modular shelf that can be easily moved. It's even got tiny wheels!

03 Each TV has its own IP address, connects to the Node.js server on Johanna's laptop, and looks out for incoming events such as video feeds.

Stockholm-based Johanna is a self-taught programmer and maker. She was a web developer and digital producer at companies such as Acne and B-Reel. Four years ago she set up on her own, specialising in conceptual art and with a specific aim of working collaboratively across different media and learning new things. It led to her ongoing interest in Raspberry Pi.

"I started using Raspberry Pi [because] I wanted to find new areas where I could use my programming skills and wished to go outside the traditional computer screen with my output, bringing in more physical and real-world objects in my work," explains Johanna. "Raspberry Pi has been the ideal solution to use in most of my interactive art installations to control the hardware."

Superstar VJ

Those with designs on becoming video DJs themselves will need a Mac and software that supports video output for Syphon (**magpi.cc/FVJQXh**), an open-source client that works with lots of video streaming programs.

Johanna's technical expertise is also a critical element. Behind the scenes of her TV Wall, she's busy controlling which bits of video footage appear on which TV and when, via the web interface she built. She can modify each Raspberry Pi's settings, including which part of each video to crop out. With such audio-visual control at her command, it's no wonder she reminds us of a modern-day kind of Oz. M

Telepresence Hand for Hazardous Areas

Manipulate objects at a distance without using space wizard powers, thanks to this remote-controlled robo-arm. **Rob Zwetsloot** tries it on for size

MAKER

Andrew Loeliger

A fourth-year student at the University of Strathclyde studying Electronic and Electrical Engineering.

▼ There's currently no wrist movement on the glove, but this can be easily added

There's a scene at the start of eighties classic *Short Circuit* where Steve Guttenberg is hiding away in a lab, programming a robot hand that is playing the piano. It's quite quaint by today's robotic standards (and was probably just a puppet at the time), which is only more apparent when viewing Andrew Loeliger's university project from the last year.

"I set out to provide a solution to the issue of first line responders operating in dangerous areas," Andrew explains. "Bomb disposal sites, biohazardous areas, and nuclear hot zones are all crucial areas where human intervention would be required but could be potentially life-threatening. The aim of the project was to develop an extremely low-cost robotic hand that can operate in hazardous areas and perform dextrous tasks while being controlled and viewed remotely. The visual feedback provided by the system allows the user to control every movement of the hand as if they were there."

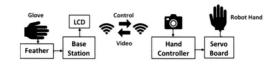

Telepresence Hand for Hazardous Areas
Block Diagram

▲ Here's the breakdown of how the system works

The user wears a glove that's connected to a 'base station', which also has a display. The display shows the pictures from a wirelessly connected camera, which is part of the remote robot hand system. The glove has a series of sensors to record how the fingers and hand are moved, and that is then relayed to the hand controller.

Hand print

"The bulk of the robot hand comprises 3D-printed components," Andrew tells us. "The design files for the 3D-printed components were sourced from InMoov. The 3D-printed robot hand is designed to hold all of the servos needed to mimic user movements made in the glove remotely. To control the robot hand, there is a Raspberry Pi Zero W which takes the received values from the base station and sends them to the servo driver board to move the servos. There is one servo for each finger and each finger is moved via a braided fishing line pulley system within the finger."

Robotic mimicry

The hand works very well, according to Andrew: "The project worked as hoped and is simple to use, as the user simply needs to put the glove

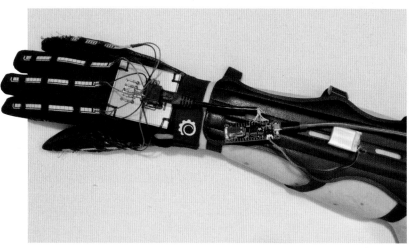

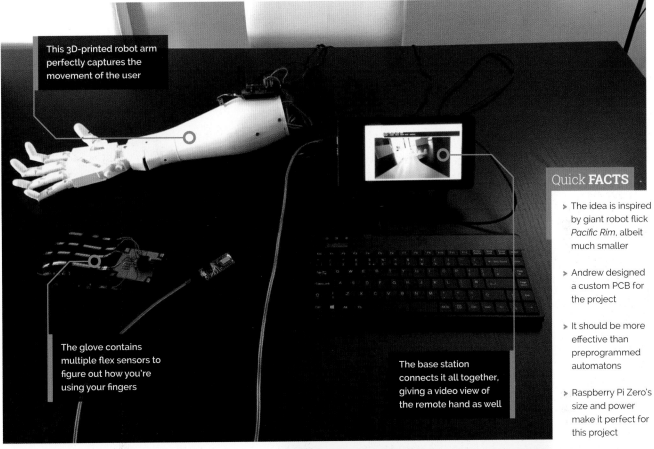

This 3D-printed robot arm perfectly captures the movement of the user

The glove contains multiple flex sensors to figure out how you're using your fingers

The base station connects it all together, giving a video view of the remote hand as well

Quick **FACTS**

> The idea is inspired by giant robot flick *Pacific Rim*, albeit much smaller

> Andrew designed a custom PCB for the project

> It should be more effective than preprogrammed automatons

> Raspberry Pi Zero's size and power make it perfect for this project

> The wrist-turning servo is actually already in place

❝ The user simply needs to put the glove on and move their hand to make the robot hand work ❞

on and move their hand to make the robot hand work. One of the really nice features is that there is low latency between the glove and robot hand movements, which means the hand truly mimics the user's finger movements."

Andrew has plans to improve the design with a new version that makes use of Bluetooth Low Energy for the glow, along with sensors so the hand can turn at the wrist, and perhaps a haptic feedback system as well.

There is one final use case that Andrew has found: "Although I originally considered this project for hazardous areas, interest has been shown in it from a prosthetics point of view." ◾

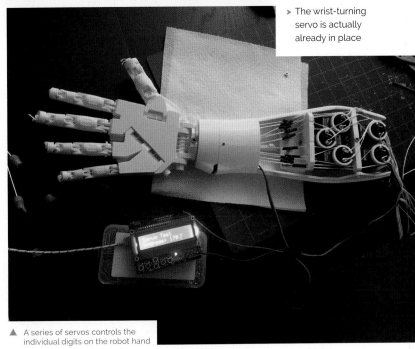

▲ A series of servos controls the individual digits on the robot hand

Pi VizuWall

This Raspberry Pi cluster has moving parts and a serious goal to help train programmers. **Phil King** finds out more

MAKER

Matt Trask

Matt is a Researcher and Chief Engineer at the FAU Machine Perception and Cognitive Robotics Lab and is grateful to the FAU Office of Undergraduate Research and Inquiry for its financial support for this project.

magpi.cc/AerxTt

Mounted on a clear acrylic plate, twelve Raspberry Pi boards suddenly spring into life, moving outwards, as if waving to the attendees at Maker Faire Miami. Not just a cool effect, the movement is proportional to each Raspberry Pi's level of activity in a parallel computing cluster. This is Pi VizuWall, a project created by long-time computer engineer Matt Trask during his degree course at Florida Atlantic University (FAU), while doing research into a new class of supercomputer systems.

"When I am successful (heh, nearly said 'if' there…), it will obsolete MPI [Message Passing Interface] as the main means of programming distributed compute clusters," explains Matt. "This means that my variant of the Beowulf architecture will function as a distributed symmetric multiprocessing system that appears to be a single unified system that is the sum of all RAM and all cores in the cluster: a virtual mainframe computer. Perhaps the solution to the so-called Ninja Gap?"

Matt is referring to the difficulty of enabling computer science students to obtain enough early experience programming parallel computing systems to become industry-proficient. Hence his motivation for building a low-cost cluster system with Raspberry Pi boards, in order to drive down the entry-level costs.

Moving parts

Matt reveals how Pi VizuWall works: "Each node is capable of moving through about 90 degrees under software control because a small electric servo motor is embedded in the hinging mechanism. The acrylic parts are laser-cut, and the hinge parts have been 3D-printed for this prototype."

While the original concept was to also use LEDs to edge-light the acrylic plate and change the colours to indicate CPU usage, Matt says the idea

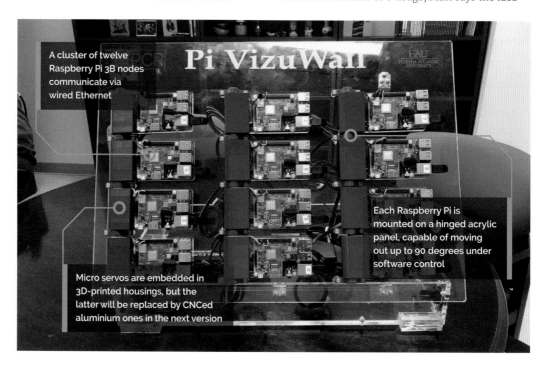

A cluster of twelve Raspberry Pi 3B nodes communicate via wired Ethernet

Each Raspberry Pi is mounted on a hinged acrylic panel, capable of moving out up to 90 degrees under software control

Micro servos are embedded in 3D-printed housings, but the latter will be replaced by CNCed aluminium ones in the next version

Hinges and housings were 3D-printed, while the acrylic panels were laser-cut

❝ The physical motion helps student programmers understand their system utilisation ❞

of the moving boards was always fundamental to the project: "I figured that the physical motion would help student programmers understand their system utilisation. And it looks cool."

Although Matt came up with the project's concept several years ago, he only started building it in late 2018. "I collaborated with Art Rozenbaum (FAU Mechanical Engineer) over the fall to develop the concept and submitted my research proposal in November," he tells us. "Art and I worked through my original concept for mounting the servos behind the board and pivoted to his concept of embedding them in the hinge mechanism, a much cleaner solution."

Currently, its Raspberry Pi boards have wired communication via a 14-port Ethernet switch, but Matt is looking into making it wireless. This will involve "evaluating whether [Raspberry] Pi's wireless LAN capability is suitable for carrying the MPI message traffic, given that the wired Ethernet has greater bandwidth."

Scale model

The original plan for Pi VizuWall was to create a 4×8 ft (1.2×2.4 m) wall with 300 Raspberry Pi boards wired as a Beowulf cluster running the MPICH implementation of MPI. "When I proposed this project to my Lab Directors at the university, they

baulked at the estimated cost of $20–25,000 and suggested a scaled-down prototype first."

Matt says some lessons have been learnt in the process of building it, including plans to replace the 3D-printed plastic motor housings – which suffered minor distortion due to heat from the servos – with CNCed aluminium. "This will [also] permit us to have finer resolution when creating the splines that engage with the shaft of the servo motor, solving the problem of occasional slippage under load that we have seen with this version."

The ultimate goal is to "create a massive piece of kinetic art to embellish the entryway to our new Lab facility at the university." ◨

▲ Ethernet cables are tucked neatly between each Raspberry Pi and its hinged panel, so as not to impede its movement

Smart Home
Herb Garden

Growing herbs using Google's Smart Home API makes for automated flavour in your cooking.
Rob Zwetsloot grabs a bunch

MAKER

Oscar Prom

Software team lead at Deeplocal. He plans, develops, and deploys a variety of full stack and cloud software systems.

magpi.cc/pNpRxP

Warning!
Mains electricity

This project uses mains power. Be careful if you plan to recreate it

I f you've ever grown herbs in your kitchen, you may have encountered some problems. Coriander flopping about everywhere. Rosemary never really regrowing. Basil growing out of control. Then you leave the house for a few days and come back to withered herbs. It's tricky! This is where something like the Smart Home Herb Garden from Oscar Prom at Deeplocal comes in handy.

"The herb garden was built for Google I/O 2019 to showcase the Smart Home API and some newly released traits on the IoT platform," Oscar explains. "We released it as a DIY project to encourage developers to use it as a jumping off point for their own Smart Home projects."

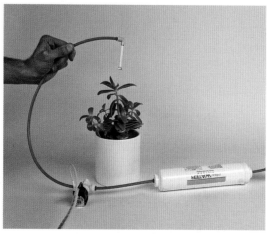

▲ Testing the mister before embedding it in a shelf is a good idea whatever project you're working on

▲ Each herb can be maintained individually, so you won't overwater one plant and underwater another

Automated gardens are all the rage now – we've had farm robots, hydroponics, and aquaponics in *The MagPi* – so scaling it down to a small herb garden seems like a logical next step. So, when Deeplocal were asked to build a Smart Home project using voice control, it's the route they decided to take.

> ❝ Three potted herbs sit under a beam that has lights and water misters ❞

Voice-activated care

The system is deceptively simple. Three potted herbs sit under a beam that has lights and water misters. There's also a humidifier on the tray that the plants sit on, and each plant can rotate to make for easier pruning and watering hard-to-reach

These decorative stones hide humidifiers and turntables for the plants

The bar that goes over each plant provides light and water when required

Each plant is monitored by the system to make sure it's getting enough water and sunlight

Quick **FACTS**

> The three herbs grown were basil, parsley, and mint

> The team unfortunately no longer have the garden

> It does have voice control, but is otherwise fully automated

> The system is written in JavaScript

> Find full build instructions here: **magpi.cc/fCPqeP**

◀ The Smart Herb Garden was created for Google I/O 2019 – here it is on display there!

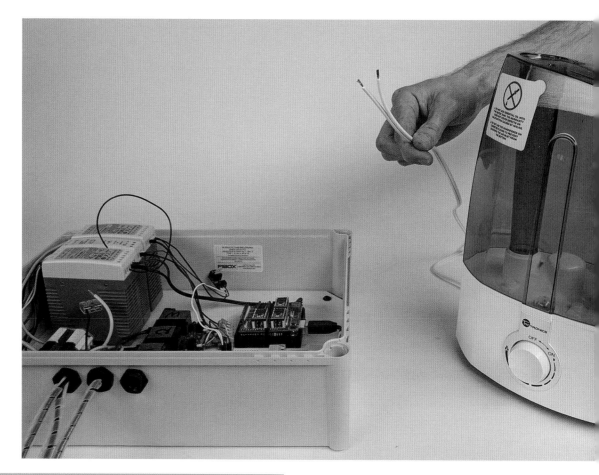

▶ A store-bought humidifier needs to be modified to work in the system

▼ The humidifiers release spooky water vapour from the rocks

Parsley   Mint

❝ It automatically rotates the plants to distribute sunlight evenly ❞

areas. There's even a special function that lets you 'spotlight' a specific plant if you want to really show off your prize parsley.

"Raspberry Pi provides a familiar and inexpensive platform to get any project off the ground," Oscar tells us. "We needed something low-power and internet-connected that could control custom hardware, and there is no dev board that hits that sweet spot better than a Raspberry Pi."

No growing pains

After having tried our own hand at growing herbs in the past, we had to ask about the project's herb-growing prowess: "It's much better than a human!" asserts Oscar. "It remembers to water the kitchen herbs without issue and automatically rotates the plants to distribute sunlight evenly. We

Building a smart herb garden

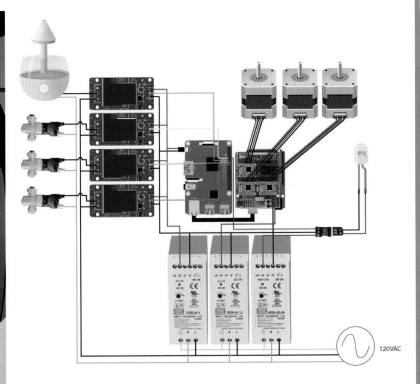

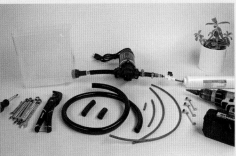

01 A water pump system needs to be properly put together to provide water for the herbs. The humidifier system is handled by an off-the-shelf humidifier controlled by a relay, so it doesn't use this system.

02 The pots use magnets to snap to a custom rotation device that is set up in three sections along the bottom of the herb garden. Each of the rotators is controlled by a simple electric motor, and they use some 3D-printed parts.

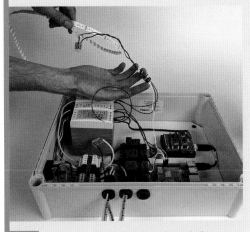

03 Lights are attached and can have their brightness controlled so that they give the perfect amount of light to the plants. This project uses white LEDs, although grow bulbs are more standard practice.

120VAC

▲ It looks a little complicated, but it's not too bad really

◄ Don't make our mistake: make sure to prune your herbs!

can even increase the brightness of the grow lights on our cloudy Pittsburgh days (read: often)."

This isn't Deeplocal's first rodeo with Raspberry Pi either, and it seems like the team specialise in amazing home improvement projects.

"We've built a [Raspberry Pi-powered], voice-controlled drink mixer and an all-in-one button that starts Netflix, turns off your lights, and orders takeout," Oscar says. "Not to mention, a lot of our prototypes are built on top of Raspberry Pi boards because we can iterate so quickly."

While the herbs in their garden grew well, they never got to use them in a meal. Maybe next time. 🅜

Marvin Go-Kart

Mark Cantrill's new mode of transport suits his go-kart-loving daughters to a tee, as **David Crookes** discovers

MAKER

Mark Cantrill

An electronics engineer, husband, and father-of-two who is usually found running the Cotswold Raspberry Jam's MicroPiNoon arena. He's known for creating the PiZ-Moto motor driver for Raspberry Pi Zero and an FPGA board for the original Raspberry Pi.

@AstroDesignsLtd

Golf and go-karting would appear to be at opposite ends of the sporting spectrum, but when Mark Cantrill swung by an ageing electric golf trolley, it set the wheels in motion for a rather unique idea.

Having originally considered using Raspberry Pi to control the trolley and eventually turn it into a Dalek or a rough-terrain garden explorer fitted with a camera, he began to ponder how it could make his life a bit easier.

"I thought I'd use it to power a go-kart and save my back from dragging my kids around the garden," he laughs. But first he had to figure how to get Raspberry Pi to control the trolley, so, after fitting a new 12V car battery, he began to look at the device's radio-controlled mode with a view to reverse-engineering it.

Go, go-kart

"The trolley could be made to go forwards, backwards, left, and right using a five-button remote control, and the fifth button would bring it to a stop," Mark explains. He soon noted that the radio receiver was a plug-in option, with a seven-pin connector joining it to the main speed-controller micro.

"I thought it would be possible to remove the radio receiver and replace it with Raspberry Pi," he says. "Since the motor controller on the golf trolley was 5V, the 3.3V outputs from Raspberry Pi needed level-shifting, which I achieved by using a 2N2222 NPN transistor. Five transistor-based inverting level shifters were quickly assembled onto a ProtoZero board."

At this point, Mark began to write the software and he was able to call upon his experience of running MicroPiNoon robots at the Cotswold Raspberry Jam. "The golf trolley had two motors, two wheels, and a front stabiliser, so it was essentially very similar to the MicroPiNoon robots, just a bit bigger.

"All I needed to do was replace the functions that usually control the EduKit 3-compatible PiZ-Moto pHAT with functions to drive the 5-bit output

The robot features the golf trolley's original motor controller, which has a handy on-board 5V power supply

Mark is building a new casing for the towing robot, with an extra wheel fitted beneath the base

Raspberry Pi is connected to a ProtoZero on which five-transistor-based inverting level shifters were assembled

Jessica and Ruth love being pulled around the garden by Marvin, with dad Mark at the controls

> # " I thought I'd save my back from dragging my kids around the garden "

that goes to the golf trolley's motor controller via the level shifter. I was able to reuse my existing code to do this."

Be in control

Mark took the enhanced prototype golf trolley to Raspberry Fields last year and allowed people to play around with it. At this point, the trolley wasn't attached to a go-kart, but it highlighted a particular problem: the PlayStation 3 controller Mark was using – a model made by Rock Candy – would shut down after a few minutes of no activity.

"It would then search for something to communicate with, but this wouldn't necessarily be the last thing it was talking to," Mark says. "It meant that it would sometimes find another robot – on one occasion it hooked up with a device called X-Bot, sending it crashing off a table."

The controller problem remains unresolved and Mark says he also needs to work on better acceleration controls. "I'm also working out how to get the controller to slow down gradually instead of stopping abruptly," he adds. But when everything is working well, Marvin is a joy. The robot can whizz around the garden with a go-kart in tow, to the great excitement of his daughters. "They love it," Mark concludes.

Quick **FACTS**

> Marvin has a Raspberry Pi Model B+

> A PlayStation 3 gamepad provides wireless control

> The interface was built on a ProtoZero board

> The project cost about £110

> Mark tried without joy to remotely control the trolley's horn

The golf trolley's radio receiver was reverse-engineered so Marvin could replicate the instructions it issues using Raspberry Pi instead

Mark is looking to securely mount the motor controller and Raspberry Pi to the towing robot and make the casing more weather-proof

Tortoise Fridge

Thanks to Raspberry Pi-controlled heating, Stefan Wollner's tortoise enjoys a cosy hibernation year after year, as **Rosie Hattersley** discovers

MAKER

Stefan Wollner

Stefan works for a large insurance company. He loves technology and taught himself programming.

Warning!
Mains electricity

This project uses mains power. Be careful if you plan to recreate it

▼ A house move prompted Stefan Wollner to consider how to adapt the new surroundings for his tortoise's hibernation

Seasonal temperature shifts can be a challenge for all living creatures. We humans can simply add or remove a jumper or vest, but our domestic companions don't have that option.

Tech enthusiast Stefan Wollner encountered exactly this issue when he moved home in 2015 and realised the basement was too warm for his pet tortoise, Pumba. Stefan had planned for the tortoise to spend the winter in a refrigerator – a common ploy for tortoise owners. However, the temperature of a standard refrigerator can't be constantly monitored and the internal thermostat only provides a limited degree of temperature control. The warmest and coldest settings of an average fridge vary just three degrees Celsius.

In a flash of inspiration, Stefan realised that he could use a Raspberry Pi and a temperature sensor to keep watch on Pumba's environment. For just a few euros, Stefan bought a DS18B20 temperature sensor to place inside the refrigerator. The system checks the temperature once a minute and records the results in a database, which it then displays as a real-time graph. The sensor's findings are then acted upon by a cooling compressor connected to a 12 V-to-230 V switching relay, operated by Raspberry Pi.

Two small Raspberry Pi-powered displays visually confirm everything is working as it should. One two-line display shows the fridge's current temperature and the temperature of the basement; the other displays the time the refrigerator was last opened, and the current status of the cooling compressor.

▶ The heart of the kitchen installation is a first-generation Raspberry Pi B

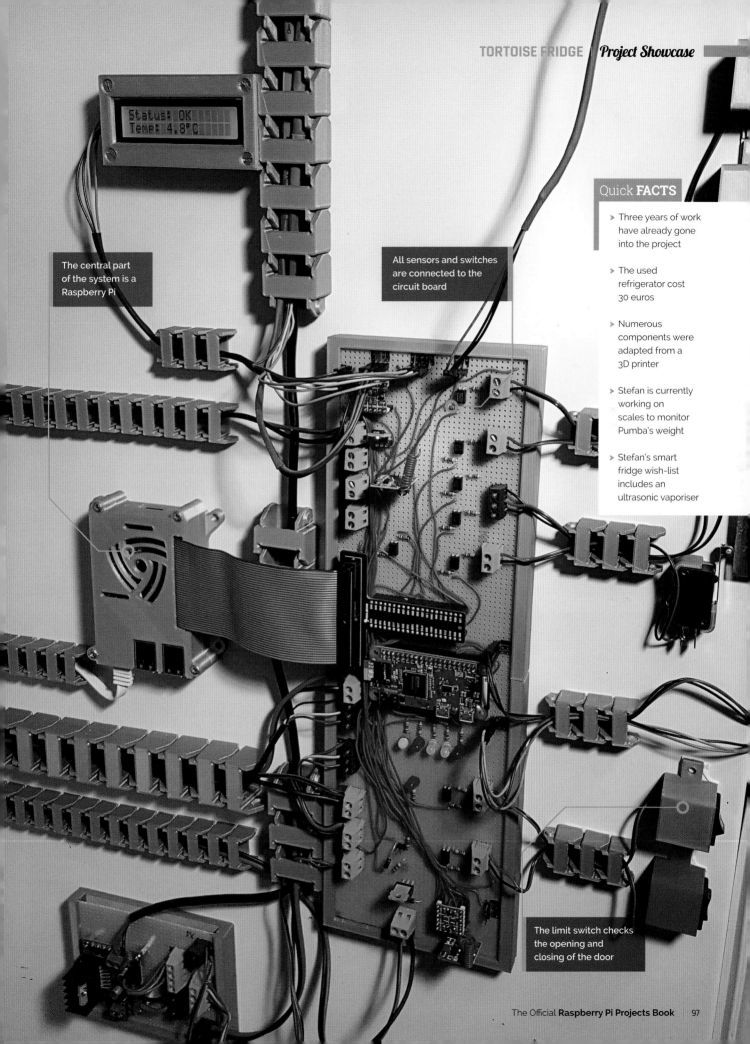

Status: OK
Temp: 4.8°C

The central part of the system is a Raspberry Pi

All sensors and switches are connected to the circuit board

Quick **FACTS**

> Three years of work have already gone into the project

> The used refrigerator cost 30 euros

> Numerous components were adapted from a 3D printer

> Stefan is currently working on scales to monitor Pumba's weight

> Stefan's smart fridge wish-list includes an ultrasonic vaporiser

The limit switch checks the opening and closing of the door

▲ A T-Cobbler Plus breakout board is used to wire the other electronics to Raspberry Pi

▶ Pumba enjoys chomping on flowers

▼ Prototyping the project on a desk

Whenever someone visits the basement, the backlight of both displays is activated by a PIR motion sensor. There's no need for Stefan to visit the basement to check up, though: another Raspberry Pi shows this data on two additional LCD displays in the kitchen. If Stefan's busy preparing food when he wants an update on Pumba's environment, he simply presses a button on the kitchen display and gets an audible update.

❝ Next up is an ultrasonic vaporiser to regulate the air humidity ❞

A breath of fresh air

Temperature control wasn't the only issue facing Stefan and his beloved tortoise. He also had to consider its oxygen supply. Whenever Stefan was at home, the fridge door would be opened regularly, refreshing the oxygen supply, but work and socialising often took him away from home. Stefan wasn't prepared to risk the unthinkable and let the oxygen supply run out. Instead, he created an automatic door-opening system with a linear motor that opens and closes the refrigerator door at preset times and checks the door has been closed properly.

The (by now) very smart fridge automatically issues error messages, warnings, and notices, keeping Stefan abreast of any issues that might arise with his carapace-clad companion's living arrangements. Should he need to, he can even remotely adjust the temperature and the timing of the fridge door opening, to allow more or less oxygen into the hibernation centre.

Emergency shutdown procedures

Stefan has spent many hours making the perfect hibernation environment for Pumba. The whole shebang is also safe from intruders and electricity outages. A Raspberry Zero W secures the whole setup. This has its own sensor in the refrigerator which measures the temperature and switches off the system completely when the threshold value has been reached, and sends a notification.

Stefan says, "Since I will probably never complete the work on this project, I am in the process of planning the next enhancements. Among other things I want to install in the near future are an ultrasonic vaporiser to regulate the air humidity and a scale for weight control." He also plans to automate what the tortoise is fed and when. ◾

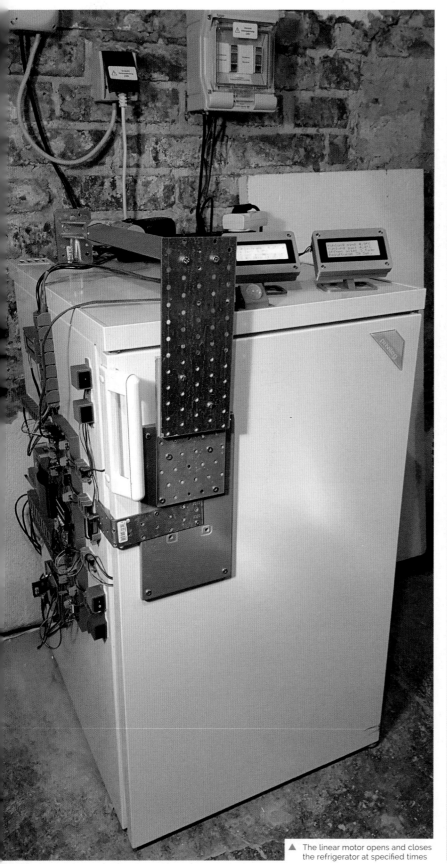

▲ The linear motor opens and closes the refrigerator at specified times

Open door policy

01 When no one's at home, the oxygen controller kicks in, automatically opening the fridge door at preset times to refresh the air supply so the resident can breathe easily.

02 A range of sensors monitor the temperature in the fridge and the levels of oxygen available. LCD displays in the basement and kitchen show the current status of the fridge and when its door was last opened.

03 Oblivious to his owner's Herculean efforts, Pumba luxuriates in a den of fresh leaves, precisely temperature-controlled and with ample fresh air and food.

Immersive Bar

Pour Sense HAT 'ingredients' to mix virtual cocktails in this fascinating blend of the physical and digital worlds. **Phil King** is shaken and stirred

MAKER

Aditi Vijay, Laurin Kraan, Minatsu Homma

Aditi, Laurin, and Minatsu are students at Copenhagen Institute of Interaction Design (CIID). Their previous backgrounds were in graphic design, graphical user interfaces, and digital consulting.

magpi.cc/TjuzKV

Mixed reality (MR) is the merging of the real and virtual worlds to produce new environments where physical and digital objects co-exist and interact in real-time. And what better way to illustrate the concept than by mixing virtual cocktails with ingredients represented by Raspberry Pi boards equipped with Sense HATs?

This was the idea of three students at Copenhagen Institute of Interaction Design (CIID) when tasked, during an 'Immersive Experiences' course, with creating a joyful mixed reality experience which offers a recurring element of surprise.

"Emerging technologies such as mixed reality are increasingly finding their way into everyday life," explains Aditi Vijay. During a brainstorming session, she and fellow students Laurin Kraan and Minatsu Homma came up with the concept for the Immersive Bar.

In this interactive game, the player chooses from three ingredients, represented by SenseHAT-equipped Raspberry Pi boards, and pours them into a cocktail shaker. "The ingredients are unknown, but by trial and error it is possible to find out what

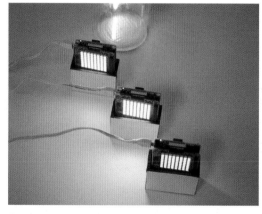

▲ Three Sense HATs and Raspberry Pi boards represent the cocktail 'ingredients' to be poured

they are," says Laurin. "In addition, the colours of the Sense HATs give a further indication of what the ingredients are."

Pour me a drink

When an ingredient is being poured, the Sense HAT's accelerometer detects the angle, while its LED matrix display enhances the visual effect with moving particles. An LED strip – controlled by a fourth Raspberry Pi – in the shaker visualises the liquid level. Once the shaker is full, a laptop computer – running Unity – mixes the cocktail and 'drinks' it before giving a rating. The whole setup is recorded by a camera and displayed on a screen.

"In total, seven different drinks can be mixed," reveals Laurin, "from pure vodka to exotic cocktails such as the Astronaut." The rating is based on two factors – 1. The combination of the ingredients: while some of them complement each other very well, others lead to a nauseous creation. 2. The balance of the ingredients: is the ratio between alcoholic and non-alcoholic ingredients well balanced?"

The project took the trio just five days to create. "The most difficult part was to turn our concept into a prototype within such a short time," recalls Minatsu. "We iterated fast and explored different digital and physical feedback to create a smooth

Drink: Astronaut
Score: 7/10

▲ The cocktail is rated by the computer and a score awarded, which is shown on the screen

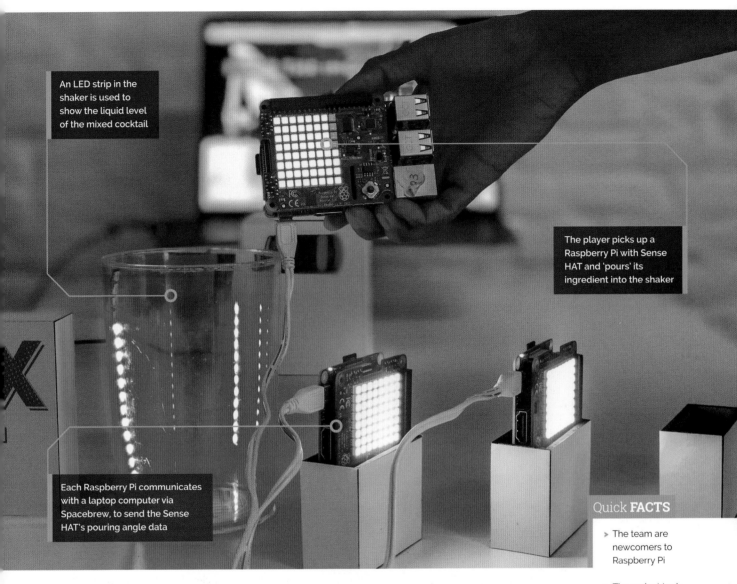

An LED strip in the shaker is used to show the liquid level of the mixed cocktail

The player picks up a Raspberry Pi with Sense HAT and 'pours' its ingredient into the shaker

Each Raspberry Pi communicates with a laptop computer via Spacebrew, to send the Sense HAT's pouring angle data

Quick **FACTS**

> The team are newcomers to Raspberry Pi

> The project took five days to design and create

> Hardware items communicate wirelessly via Spacebrew

> Seven different drinks can be mixed

> The ingredients are a mystery to the player

and all-round experience. For example, we initially wanted to use a water pump to raise the liquid level inside the shaker while virtually pouring the ingredients. It turned out that the pumps we had were not strong enough and we had to switch to an LED strip for the liquid level indication."

Student demonstration

When demonstrated to other students, faculty, and guests at CIID, the Immersive Bar project was very well received and soon had people queueing up to play. You can watch a video of the cocktail-mixing fun at **magpi.cc/MZbAXC**.

"The merging boundaries between the physical and digital worlds left the audience amazed," says Laurin. "The demonstration soon turned into a competitive experience, in which the players tried to figure out how to achieve a better score than their friends."

" The merging boundaries between the physical and digital worlds left the audience amazed "

The project was a big hit with fellow students and guests when demonstrated at CIID

Get Started with Electronics

114

126

130

134

Get Started with Electronics

MAKE
WITH
CODE

If you have bought a Raspberry Pi to learn how to code and hack new electronic gizmos, then you have made a great choice. Get started with this guide...

If you have just got a brand new shiny Raspberry Pi, you may have plugged it in and got it working. You may have played a few of the games or tried out the applications, or maybe you've loaded one of the programming tools and then looked at it wondering what to do next. If you haven't done any programming before, we have wisdom for you here. In the next few pages you'll find answers and probably some questions, but then some more answers. Before you know it you will be a coding, hacking ninja.

> ❝ If you have never done any programming before, this may appear a bit daunting, but it's quite easy really ❞

If you're new to programming, this may appear a bit daunting, but it's quite easy really – you've just got to get stuck in and start with some simple things that will get you results. One of the programming languages that is supplied with the Raspberry Pi is Python. It's very easy to get started with Python and it can be used to program many of the add-ons that are available for the Raspberry Pi – so this is going to be very useful to learn. We can get you up and coding in 30 seconds flat; just read on…

MAKE AND RUN A PROGRAM

To get started with coding is really easy. Coding is just giving the Raspberry Pi an instruction to do something. The only thing you need to know is what language to talk to it in. In this case we are going to talk in Python 3. We will need to write our instructions somewhere so, to start, open a Terminal window – click the icon in the top left of the screen which is a grey box with a blue bar across the top. This opens a black window with a prompt: **pi@raspberrypi:~ $** in green. If you type **python3** and hit **ENTER**, Python starts and you will see the triple chevron prompt: **>>>**. Now type **print("Hello")** and hit **ENTER**. Bosh! Your first program. You have instructed your Raspberry Pi to print the word 'Hello' and, all being well, it has obeyed.

Entering programs like this is not going to be very useful most of the time, so now let's look at an app we can write and save a program with. Go to the desktop menu (click the Raspberry Pi logo in the top left of the screen) and in the Programming section, select Thonny (which should then open by default in its Simple mode). Try typing in the following program in the Thonny editor and save it, then run it by clicking the Run button. The output will be displayed in the Shell frame below the program editor.

Mark Vanstone

Educational software author from the nineties, author of the ArcVenture series, disappeared into the corporate software wasteland. Rescued by the Raspberry Pi!

magpi.cc/YiZnxL
@mindexplorers

mwc1.py

```python
001. import random
002.
003. correct = False
004. r = random.randint(1,10)
005. c = 0
006. while correct == False:
007.     n = input("Guess my number between 1 and 10: ")
008.     c = c + 1
009.     if int(n) == r:
010.         correct = True
011.     else:
012.         if int(n) > r:
013.             print("Sorry, my number is lower. Try again.")
014.         else:
015.             print("Sorry, my number is higher. Try again")
016.
017. print("Well done. The correct answer was " + str(r) + ".
     You got it in " + str(c) + " tries.")
```

Language: Python
magpi.cc/HNJhhd

LEARN TO CODE

Coding involves using several elements. Let's take a look at a range of them and how they work

Import modules

Define data

updateAnswer()

Make new answer string

startGuessing()

While not finished, input new guess

Player wins

Player loses

End

▲ After initialising the data the program needs, we loop round, updating the answer variable until the player guesses all the letters or they run out of tries

In the last program, we get input from the keyboard and output text and numbers by using the **print()** function. We have also used a condition structure in the form of **if** and **else**. Python is very particular about how you indent the code with spaces (four per indent level); this shows that the indented code is inside another structure. In the last program, everything indented after the **while** statement will be part of that loop. Let's take a look at a few more coding techniques.

01 Using lists

We are going to write a Hangman-style game, where we start with a secret word and the player has to guess it, letter by letter. If correct, we show them where that letter appears in the word. They are allowed ten wrong attempts before the game ends. See the **mwc2.py** code to follow along. First, we make a list of words to choose from. A list is defined in Python using square brackets, like: **list = ["a","b","c"]**. We'll call our list **WORDLIST**. In this case we're writing the list name in upper case to show that it is a constant, i.e. It is not going to change throughout the program.

02 Pick a word

We can pick a word from our list using the **random** module. We import the module at the top of our code, then we can use the **random.choice()** function to get a word and store it in a variable:

theWord. When we call a function that is inside a module, we use a full stop between the module name and the function name. Next, we want to get the player to start guessing what the word is. If we look at the bottom of the code, we can see that we call a function called **startGuessing()**. This is our own function that we need to define.

03 Defining functions

Each time we call a function, the code inside it runs. Sometimes functions have outputs, like our function **updateAnswer()** that returns the variable **result**. One of the rules of Python is that you must define a function before you call it, so we will need to define our **startGuessing()** function near the top of the code. To do so, we write **def** and then the name of the function, followed by brackets and a colon. If we want to pass variables as inputs into a function, we can add them inside the brackets.

04 Getting loopy

Now for the code in our **startGuessing()** function. We set the number of tries and dashes, one for each letter of the secret word, then we go into a loop. The code says: "While the player still has some tries left and the answer we have is not the secret word, run the following code." In our loop, we print the answer we have so far and how

mwc2.py

Language: **Python**
magpi.cc/RqQdhR

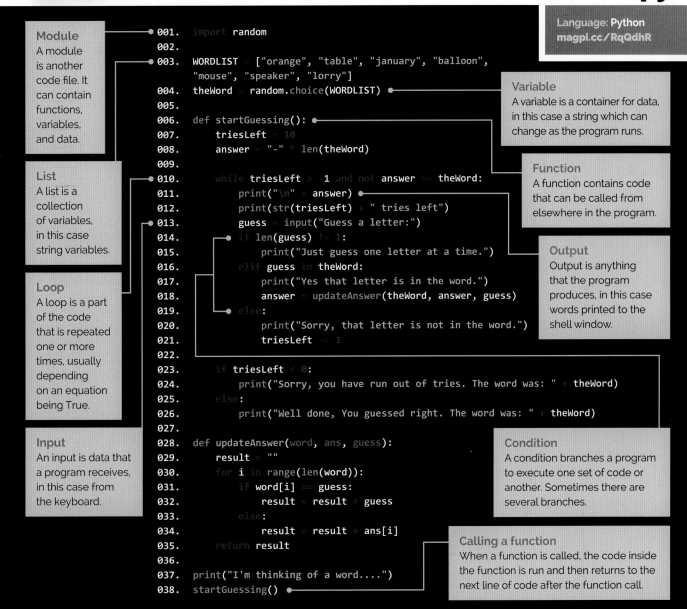

Module
A module is another code file. It can contain functions, variables, and data.

List
A list is a collection of variables, in this case string variables.

Loop
A loop is a part of the code that is repeated one or more times, usually depending on an equation being True.

Input
An input is data that a program receives, in this case from the keyboard.

Variable
A variable is a container for data, in this case a string which can change as the program runs.

Function
A function contains code that can be called from elsewhere in the program.

Output
Output is anything that the program produces, in this case words printed to the shell window.

Condition
A condition branches a program to execute one set of code or another. Sometimes there are several branches.

Calling a function
When a function is called, the code inside the function is run and then returns to the next line of code after the function call.

```python
001.   import random
002.
003.   WORDLIST = ["orange", "table", "january", "balloon",
       "mouse", "speaker", "lorry"]
004.   theWord = random.choice(WORDLIST)
005.
006.   def startGuessing():
007.       triesLeft = 10
008.       answer = "-" * len(theWord)
009.
010.       while triesLeft > -1 and not answer == theWord:
011.           print("\n" + answer)
012.           print(str(triesLeft) + " tries left")
013.           guess = input("Guess a letter:")
014.           if len(guess) != 1:
015.               print("Just guess one letter at a time.")
016.           elif guess in theWord:
017.               print("Yes that letter is in the word.")
018.               answer = updateAnswer(theWord, answer, guess)
019.           else:
020.               print("Sorry, that letter is not in the word.")
021.               triesLeft -= 1
022.
023.       if triesLeft < 0:
024.           print("Sorry, you have run out of tries. The word was: " + theWord)
025.       else:
026.           print("Well done, You guessed right. The word was: " + theWord)
027.
028.   def updateAnswer(word, ans, guess):
029.       result = ""
030.       for i in range(len(word)):
031.           if word[i] == guess:
032.               result = result + guess
033.           else:
034.               result = result + ans[i]
035.       return result
036.
037.   print("I'm thinking of a word....")
038.   startGuessing()
```

many tries the player has left. Then we use an **if**, **elif**, **else** condition structure to respond to the player, depending on what they typed.

guessed all the right letters in our word or they have got it wrong ten times, the program will drop out of the loop to reach the final part of the function.

05 Changing the answer

If the player guesses a letter correctly, we call another function: **updateAnswer()**. This uses a **for** loop to add the correct letters into our answer variable, then return that string (a variable containing letters/characters rather than a numeric value). This then becomes the answer variable that we print at the start of each loop in the **startGuessing()** function. When the player has

06 Win or lose

We have an **if** and **else** structure to print congratulations, or let the player know they've run out of tries. A few of the functions are used with variables: **len()** finds the length of a string, and **str()** converts a number variable into a string so it can be added to the start or the end of another string. After the function is complete, it returns to where it was called, which is the end of the program.

CONTROL THINGS WITH CODE

Now we've got the hang of the coding, let's put it to work with some electronic components

In the previous example, we imported a module into our code and used functions from it. If we want to control electronics from code, there is a very useful module available called **gpiozero**. GPIO stands for 'general-purpose input/output' and the line of double pins on one side of the Raspberry Pi are called GPIO pins. For details about the labels of all the pins, see **pinout.xyz**. If we connect electrical components to these GPIO pins, we can use the **gpiozero** module to make things happen. When we import a module, there are often many different functions inside. We can also make new coding 'objects' with them. Objects are like variables but have their own set of functions and properties inside them that we can call or change, and we do that using the same dot notation (full stop) that we did with the **random** module. For more details on using coding objects (called object-oriented programming or OOP), see issue 54 of *The MagPi*: **magpi.cc/54**.

You'll Need

> A breadboard
> **magpi.cc/NtjSiy**
>
> An LED
> **magpi.cc/WBVPxG**
>
> A resistor
> **magpi.cc/iDTFag**
>
> 2 × male-to-female jumper wires
> **magpi.cc/OkMyVX**

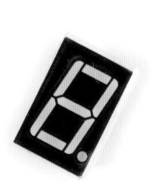

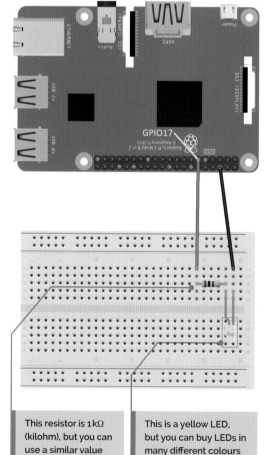

GPIO17

This resistor is 1 kΩ (kilohm), but you can use a similar value

This is a yellow LED, but you can buy LEDs in many different colours

01 The breadboard

Breadboards come in various sizes, but they all work in the same way. If there are tracks marked red and black/blue (and/or + and −) on the long edges of the board, these are for power and are connected along the length – although sometimes divided into sections. The matrix of holes which make up the main part of the board are connected in lines the other way (vertically in the diagram). There is usually a break in the centre of the board so that the two sides are not connected.

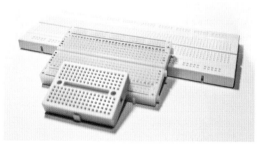

02 Light-emitting diode (LED)

An LED is a bit like a bulb in that if you apply electricity to it in the right way, it lights up. An LED is also a diode, which means that the electricity needs to flow in the correct direction or it will not light. When connecting an LED to the Raspberry

Pi, we need to add a resistor to the circuit, as most LEDs will burn out if we connect them directly to the main power output. In the example, we are using a 1 kΩ resistor, but it's fine to use another similar value. To light an LED using **gpiozero**, assemble the components as in the diagram, then write a Python program: start with **from gpiozero import LED**, then create an LED object with **led = LED(17)**, and finally type **led.on()** to light the LED.

▲ You can control the state of the LED using the gpiozero module in Python

03 A resistor

Resistors do what their name suggests: they resist the flow of electricity (current). Some components need to have a certain amount of current in order to operate correctly. Resistors enable us to set the current or voltage to a suitable level for the other components we are using. There are many different types of resistors, with different resistance values. The resistance value can be read from the pattern of coloured stripes on the resistor. You can also get variable resistors, which are known as potentiometers.

04 Jumper wires

We need to connect our components to the Raspberry Pi GPIO pins. For this we use jumper wires. The ones we will be using have a female connector at one end, to go on the GPIO pins, and a male connector to go in the holes of the breadboard. You can also get jumper wires with both male connectors or both female connectors, for different situations. They can be bought in strips all joined together, sometimes known as 'jumper jerky'.

05 Potentiometer

A potentiometer is a variable resistor. It usually has a turning spindle that changes the resistance from one value to another, quite often from no resistance to full resistance (no electricity flowing). A potentiometer can provide us with a variable output voltage which we can measure with the GPIO, but there is a slight problem. The potentiometer provides an 'analogue' output (varies continuously between values) and the GPIO inputs are only digital, i.e. on or off. So we need another component: an analogue-to-digital converter (ADC).

06 Analogue-to-digital converter

ADC components come in various forms, but the one we have in this example is called an MCP3008. It's an integrated circuit (IC), meaning that it's a box with some circuitry inside it. We don't need to know what is inside it – we just need to know what to connect to each of the legs of the IC. We will need to wire up several of the legs to GPIO pins and provide the IC with power; when we've done that, we can connect the potentiometer to the IC and then read values in showing the position of the potentiometer spindle using the **gpiozero** module. We'll cover the code in the next section.

Electronics guide

For more details about using electronic components with the Raspberry Pi, check out our Electronics Starter Guide in issue 64 of *The MagPi*.

magpi.cc/64

PONG WITH POTS

Now to put what we have covered to the test: we will make a retro game and control it with our own homemade controllers

You'll Need

Breadboard
magpi.cc/vhZkGh

6 × male-to-female
jumper wires &
10 × male-to-male
jumper wires
magpi.cc/ffZNwL

2 × potentiometers
magpi.cc/oZRFEe

MCP3008
integrated circuit
magpi.cc/aCvXuo

On the previous page we talked about potentiometers and analogue-to-digital converters, and this is where we get to use them. It's a bit more complicated than lighting up an LED, but only a little. First, we are going to write a program which has two rectangular bats at each side of the screen that can be moved up and down by two players. A ball bounces backwards and forwards between the bats until one player misses the ball and the other player scores a point. That's right, you guessed it, the game is Pong and we are going to create a controller for each player from a potentiometer and wire it into the Raspberry Pi.

01 Super-fast game coding

If you have been following other coding articles in recent issues of *The MagPi*, you will know that when writing a quick game on the Raspberry Pi, Pygame Zero is your friend. We can make the basics of the game code very quickly by importing the **pgzrun** module, which holds all the Pygame Zero code. We need to call **pgzrun.go()** at the end of our code, and that's our game window sorted. As with all Pygame Zero programs, we have a **draw()** function to write the graphics to the window, and an **update()** function to update the game items between draw cycles.

02 Running the code

The listing **mwc3.py** provides all the code you need for the game to work. There is some code in the **updatePaddles()** function for the keyboard to control the paddles or bats, just in case you want to test it before making the proper controllers. We import several modules with this code. We have covered **pgzrun**, but we also need **random** so that the ball will move in a random direction each time it starts. In addition, we need **gpiozero** to deal with the input from the controllers, and we need the **math** module for calculating the direction of the ball.

03 Wiring it all up

One thing to bear in mind when connecting any electronics to a computer is that if the wires are connected in the wrong way, you may cause damage to the computer or the electronic

▲ It may seem like a lot of wires, but work through the diagram methodically to make sure they are plugged in correctly

components, so it's always a good idea to power off your Raspberry Pi before connecting anything to the GPIO pins. Follow the wiring diagram (overleaf) carefully, making sure that the jumper wires are connected to the right GPIO pins and to the correct places on the breadboard. Once you have put everything in place, it's a good idea to have another check just to make sure.

04 The MCP3008 IC

The MCP3008 converts the voltage from our potentiometers into a number with the help of the **gpiozero** module. It has eight channels for input, but we are going to just use two of them in this

case. You will see from the diagram that all the legs on the top side of the IC are connected to either GPIO pins or to power lines. There are two red connections going to the positive power track, then a black lead to the negative or ground track. Then there are four coloured wires that go to: purple – GPIO 11; green – GPIO 09; orange – GPIO 10; and blue – GPIO 08. Then there is one last connector to the ground track.

05 The inputs

All the MCP3008 pins on the bottom side of the IC are input channels. We will use the first two pins, which are channel 0 and channel 1. We

The MCP3008 straddles the centre break of the breadboard so that the pins on either side are not connected

GPIO11 GPIO09 GPIO10

GPIO08

The left-hand potentiometer's middle pin connects to pin 0, and the right one's to pin 1 of the MCP3008

MCP3008

connect the middle pins of our potentiometers to those channel pins, which will read the positions and convert that to a value in our program. If you want to know exactly what all the pins are for on this IC, you can do a web search for 'MCP3008 pinout' and that will give you descriptions of each. ICs are very useful in electronics as they mean that we can reduce how complex our circuits are and we don't need to know exactly how they work inside. They are a little bit like modules in Python.

06 Finishing the job

When you have checked that all the connections are correct, you can switch the Raspberry Pi back on and reload your program. You may want to initialise the SPI interface on your Raspberry Pi by going to the main desktop menu > Preferences > Raspberry Pi Configuration, and go to the Interfaces tab. It will work without this, but may cause a few warnings in the Thonny shell window. So, all being well, when you run your program you will have a game of Pong which can be controlled by two players with the potentiometers. If it doesn't work first time, check your code and then your wiring, and try again. Ⓜ

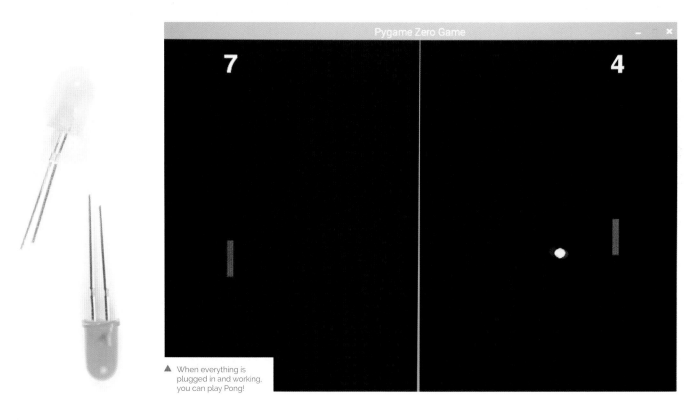

▲ When everything is plugged in and working, you can play Pong!

mwc3.py

Language: Python
magpi.cc/umUcfq

```python
001. import pgzrun
002. import random
003. from gpiozero import MCP3008
004. import math
005.
006. pot1 = MCP3008(0)
007. pot2 = MCP3008(1)
008.
009. # Set up the colours
010. BLACK = (0  ,0  ,0  )
011. WHITE = (255,255,255)
012. p1Score = p2Score = 0
013. BALLSPEED = 5
014. p1Y = 300
015. p2Y = 300
016.
017. def draw():
018.     screen.fill(BLACK)
019.     screen.draw.line((400,0),(400,600),"green")
020.     drawPaddles()
021.     drawBall()
022.     screen.draw.text(str(p1Score) , center=(105,
     40), color=WHITE, fontsize=60)
023.     screen.draw.text(str(p2Score) , center=(705,
     40), color=WHITE, fontsize=60)
024.
025. def update():
026.     updatePaddles()
027.     updateBall()
028.
029. def init():
030.     global ballX, ballY, ballDirX, ballDirY
031.     ballX = 400
032.     ballY = 300
033.     a = random.randint(10, 350)
034.     while (a > 80 and a < 100) or (a > 260 and a <
     280):
035.         a = random.randint(10, 350)
036.     ballDirX = math.cos(math.radians(a))
037.     ballDirY = math.sin(math.radians(a))
038.
039. def drawPaddles():
040.     global p1Y, p2Y
041.     p1rect = Rect((100, p1Y-30), (10, 60))
042.     p2rect = Rect((700, p2Y-30), (10, 60))
043.     screen.draw.filled_rect(p1rect, "red")
044.     screen.draw.filled_rect(p2rect, "red")
045.
046. def updatePaddles():
047.     global p1Y, p2Y
048.
049.     p1Y = (pot1.value * 540) +30
050.     p2Y = (
     pot2.value * 540) +30
051.
052.     if keyboard.up:
053.         if p2Y > 30:
054.             p2Y -= 2
055.     if keyboard.down:
056.         if p2Y < 570:
057.             p2Y += 2
058.     if keyboard.w:
059.         if p1Y > 30:
060.             p1Y -= 2
061.     if keyboard.s:
062.         if p1Y < 570:
063.             p1Y += 2
064.
065. def updateBall():
066.     global ballX, ballY, ballDirX, ballDirY,
     p1Score, p2Score
067.     ballX += ballDirX*BALLSPEED
068.     ballY += ballDirY*BALLSPEED
069.     ballRect = Rect((ballX-4,ballY-4),(8,8))
070.     p1rect = Rect((100, p1Y-30), (10, 60))
071.     p2rect = Rect((700, p2Y-30), (10, 60))
072.     if checkCollide(ballRect, p1rect) or
     checkCollide(ballRect, p2rect):
073.         ballDirX *= -1
074.     if ballY < 4 or ballY > 596:
075.         ballDirY *= -1
076.     if ballX < 0:
077.         p2Score += 1
078.         init()
079.     if ballX > 800:
080.         p1Score += 1
081.         init()
082.
083.
084. def checkCollide(r1,r2):
085.     return (
086.         r1.x < r2.x + r2.w and
087.         r1.y < r2.y + r2.h and
088.         r1.x + r1.w > r2.x and
089.         r1.y + r1.h > r2.y
090.     )
091.
092. def drawBall():
093.     screen.draw.filled_circle((ballX, ballY), 8,
     "white")
094.     pass
095.
096. init()
097. pgzrun.go()
```

Electronics
with GPIO Zero 1.5

Take an off-the-shelf battery-powered door-bell and hack it to add new features, add a camera, and use a tonal buzzer to create your own jingle

MAKER

Ben Nuttall

Ben is the creator of GPIO Zero and piwheels, and is the Raspberry Pi Foundation's resident Python expert.

@ben_nuttall

You'll Need

> Raspberry Pi Zero or 3A+
> **magpi.cc/wHwfNX**

> Raspberry Pi Camera Module
> **magpi.cc/jbKzbf**

> Jam HAT, or buzzer and transistor
> **magpi.cc/jMxZCU**

> Wilko Portable Standard Door Chime
> **magpi.cc/pfUhqo**

▶ A £10 Wilko door-bell comes with a remote battery-powered chime unit

T he GPIO Zero library for Python now includes support for making tunes with buzzers, so it's the perfect opportunity to hack yourself a new door-bell jingle.

01 Upgrade GPIO Zero

GPIO Zero v1.5 recently came out. It's got some new features we're going to use in this tutorial, so make sure you upgrade! Check that your Pi's online, open a Terminal, and type:

```
sudo apt update
sudo apt install python3-gpiozero
```

That will bring in the latest GPIO Zero, and will install python3-colorzero too, as that's a new dependency of GPIO Zero.

You can check what version you have by typing:

```
apt-cache policy python3-gpiozero
```

This will also tell you if there's a new version available.

02 Tonal buzzer

One of the new device classes in GPIO Zero is `TonalBuzzer`, which allows you to play particular tones by setting the PWM frequency. You can play a sequence of notes to make a tune, or you can make interesting sound effects like a police siren by cycling through frequency ranges at different speeds. It's nothing like the quality you'd get from a speaker, but it's certainly possible to make discernible tones and jingles. There's a tonal buzzer on ModMyPi's new Jam HAT, or you can use a normal buzzer like the one in the CamJam kit, but you'll get better results if you use a transistor to apply 5 V to the buzzer, like they do on the Jam HAT.

03 Playing a tune

First of all, open up a Python shell or the REPL in your favourite Python editor and import the stuff you'll need, then create a `TonalBuzzer` object on the GPIO pin it's connected to:

```
from gpiozero import TonalBuzzer
from gpiozero.tones import Tone
from gpiozero.tools import sin_values
from time import sleep

tb = TonalBuzzer(20)
```

Now try playing a single note:

```
tb.play(60)
```

That will play MIDI note 60 (middle C). You'll get an 'ambiguous tone' warning, but don't worry: that's just because the `Tone` interface allows you to use MIDI notes, frequencies, or musical notation. To be more explicit, you would generally use

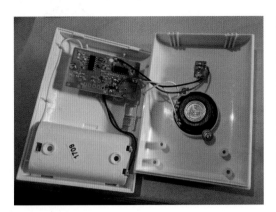

▲ Take a look inside

`Tone(midi=60)` rather than just `60` or `Tone(60)`. Type `tb.stop()` to stop it playing the note. Type `Tone(60)` into the REPL to see its three representations:

```
<Tone note='C4' midi=60 frequency=261.63Hz>
```

Try playing a scale:

```
for note in 'C4 D4 E4 F4 G4 A4 B4 C5'.
split():
    tone = Tone(note)
    print(repr(tone))
    tb.play(tone)
    sleep(0.3)
tb.stop()
```

And finally, for effect, let's try that siren noise we were telling you about:

```
tb.source = sin_values()
```

Simply passing in a series of values from the sine wave will make the buzzer continuously alter its frequency from one octave down to one octave up, achieving a siren effect. To speed it up or slow it down, alter the buzzer's `source_delay` (default 0.1):

```
tb.source_delay = 0.01
```

…or:

```
tb.source_delay = 0.5
```

04 Keyboard control

Now install the `inputs` library, which is a great utility for reading (amongst other things) key presses in real-time. Open a Terminal and type:

❝ Create a dictionary mapping each of the characters on the middle row of your keyboard to a note ❞

```
sudo pip3 install inputs
```

Return to the Python shell, import `get_key`, and create a dictionary mapping each of the characters on the middle row of your keyboard to a note:

```
from inputs import get_key

keys = {'A': 'C4', 'S': 'D4', 'D': 'E4',
'F': 'F4',
        'G': 'G4', 'H': 'A4', 'J': 'B4',
'K': 'C5'}
```

Add a loop to look for these key presses and play notes when they're pressed:

```
while True:
    events = get_key()
    for event in events:
        if event.ev_type == 'Key' and
event.code[-1] in 'ASDFGHJK':
            if event.state:
                tb.play(keys[event.code[-1]])
            else:
                tb.stop()
```

Now press any of the keys **A** to **K** and it'll play a note. You've turned your computer keyboard into a musical keyboard! Try playing a tune.

▲ Connecting the door-bell wiring to Raspberry Pi's GPIO pins

Top Tip 👍

Programming paradigms

Try out different programming styles – GPIO Zero provides different approaches including procedural, blocking, and callbacks. Read the documentation for a better understanding.

Top Tip 👍

Solder on

Be cautious when soldering – you don't want to accidentally desolder some other useful parts of the door-bell's PCB. Take your time.

05 Your own door-bell jingle

We worked out the tune for the *Pink Panther* theme. Use this, or feel free to take this opportunity to create your own, using MIDI notes or musical notation. Or you could opt for the police siren effect, but that might be more alarming than necessary when it's just an Amazon order delivery.

For *Pink Panther*, here's a function to take a list of notes and durations and play them in sequence.

```python
def play(tune):
    for note, duration in tune:
        print(note)
        tb.play(note)
        sleep(float(duration))
    tb.stop()

tune = [('C#4', 0.2), ('D4', 0.2), (None, 0.2),
    ('Eb4', 0.2), ('E4', 0.2), (None, 0.6),
    ('F#4', 0.2), ('G4', 0.2), (None, 0.6),
    ('Eb4', 0.2), ('E4', 0.2), (None, 0.2),
    ('F#4', 0.2), ('G4', 0.2), (None, 0.2),
    ('C4', 0.2), ('B4', 0.2), (None, 0.2),
    ('F#4', 0.2), ('G4', 0.2), (None, 0.2),
    ('B4', 0.2), ('Bb4', 0.5), (None, 0.6),
    ('A4', 0.2), ('G4', 0.2), ('E4', 0.2),
    ('D4', 0.2), ('E4', 0.2)]

play(tune)
```

06 Camera test

The easiest way to initially test the camera is to open a Terminal and type:

```
raspistill -k
```

This opens the camera preview until you kill it with **CTRL+C**. Now back to Python; write some simple code to trigger capturing a photo in a loop:

```python
from picamera import PiCamera
from datetime import datetime

camera = PiCamera()

while True:
    input("Press Enter to capture")
    dt = datetime.now().isoformat()
    camera.capture('/home/pi/{}.jpg'.format(dt))
```

This will take a picture when **ENTER** is pressed, and save it with the current timestamp.

07 Hacking the door-bell

Note: These instructions are for battery-powered door-bells only. DO NOT use a mains-powered door-bell.

Open up the door-bell chime by unscrewing the case to reveal the PCB inside. You'll notice a white wire with one end unattached (that's the radio receiver), red and black wires going to the battery pack, more white wires going to the speaker, and another pair of red and black wires going to the LED on the front of the case. When the door-bell is pressed and the signal is received, the LED comes on and the chime makes its sound.

Optionally, you can choose to rewire the power to the receiver so that it comes from your Raspberry Pi instead of batteries. Since it takes two AA batteries (1.5 V each), it requires ~3 V. You can power it from Raspberry Pi by connecting the black wire to GND and the red wire to 3V3 (we soldered outs straight to a 3V3 pin hole on a Raspberry Pi Zero). And of course, if you prefer your own jingle, you can disconnect the speaker.

Next, desolder the wires going to the LED and connect them to Raspberry Pi instead: red to a GPIO pin, and black to GND (you can cut jumper wires in half and solder them to the wires, and connect the female end to the pins, or even solder the ends directly Raspberry Pi Zero pin holes). It's best to test before you commit to soldering things together.

08 Testing the door-bell

Did it work? Let's see. In theory, you should be able to detect a signal on the GPIO pin you wired the door-bell to, the same way you'd detect a button being pressed. So, back to Python and GPIO Zero:

```python
from gpiozero import Button

doorbell = Button(21)

button.wait_for_press()
print("Pressed")
button.wait_for_release()
print("Released")
```

Now press the door-bell, and you should see the message 'pressed' when the door-bell is pressed, and 'released' once the chime stops playing. If it doesn't seem to be working, you'll need to check your wiring. You might need to add a resistor to make sure you detect the change in voltage when the button is pressed. We found that if we wired

▲ Connecting the door-bell wiring to Raspberry Pi's GPIO pins

ours directly to the IC (red square on the picture), the resistor in between would take effect.

09 Putting it all together

Now let's put together all the components: the door-bell, the jingle, and the camera. If you want to dedicate a Pi to live in this project, it should probably be a Pi Zero W or 3A+. With a Pi Zero, you can solder the components directly to the GPIO pin holes. You'll have easier access to the GPIO pins with the 3A+ or another full(er)-size Pi, but it will take up more space, which may be at a premium if you're trying to fit it all inside a small container (especially the original door-bell receiver enclosure).

Connect your Jam HAT or your buzzer, and solder your door-bell receiver wires and connect them to Raspberry Pi's GPIO pins. Connect the camera and find a way to mount it in place. If you've got clear glass on your front door, you can stick your Raspberry Pi behind or beside the door and have the camera facing out. Make sure all your connections are strong, and the enclosure is keeping everything in place, then connect Raspberry Pi's power supply, making sure it won't get pulled out. Attach the door-bell to the front door, and put your code together to give it a whirl! See the completed code in the **doorbell.py** listing.

10 Run at boot

The important finishing touch is to have your program run at boot. There are a good few ways of doing this, and if you want to find out some alternatives, just ask someone how they tend to do it – chances are you'll get a different response each time. Your author's preference for basic script launching like this is cron, either using crontab if you know how, or we'd recommend installing GNOME Schedule, a cron GUI:

```
sudo apt install gnome-schedule
```

doorbell.py

> Language: **Python 3**

```python
001. from gpiozero import TonalBuzzer, Button
002. from picamera import PiCamera
003. from datetime import datetime
004.
005. buzzer = TonalBuzzer(20)
006. doorbell = Button(21)
007. camera = PiCamera()
008.
009. def play(tune):
010.     for note, duration in tune:
011.         buzzer.play(note)
012.         sleep(float(duration))
013.     buzzer.stop()
014.
015. pink_panther = [
016.     ('C#4', 0.2), ('D4', 0.2),  (None, 0.2),
017.     ('Eb4', 0.2), ('E4', 0.2),  (None, 0.6),
018.     ('F#4', 0.2), ('G4', 0.2),  (None, 0.6),
019.     ('Eb4', 0.2), ('E4', 0.2),  (None, 0.2),
020.     ('F#4', 0.2), ('G4', 0.2),  (None, 0.2),
021.     ('C4', 0.2),  ('B4', 0.2),  (None, 0.2),
022.     ('F#4', 0.2), ('G4', 0.2),  (None, 0.2),
023.     ('B4', 0.2),  ('Bb4', 0.5), (None, 0.6),
024.     ('A4', 0.2),  ('G4', 0.2),  ('E4', 0.2),
025.     ('D4', 0.2),  ('E4', 0.2)
026. ]
027.
028. while True:
029.     doorbell.wait_for_press()
030.     dt = datetime.now().isoformat()
031.     camera.capture('/home/pi/{}.jpg'.format(dt))
032.     play(pink_panther)
```

Just add a new recurrent task that launches at reboot. The task should look something like this:

```
python3 /home/pi/doorbell.py &
```

11 Make it your own

So far we've added some basic features – a custom jingle and camera – but there's much more you could do. If you have smart light-bulbs, you could have it flash the house lights to give a visual indication when the door-bell rings, even when you have music playing. You could have it email, text, or tweet you. How about a push notification to your phone? If you're feeling ambitious, you could try to identify the door knocker and decide who to let in! The world is your door-bell. ◾

Teenage
Dinner Klaxon

Always yelling 'Dinner!' to a teenager or shed-dweller? Save your vocal cords by installing a tower light you can control from your phone

MAKER

PJ Evans

PJ is a writer, developer, and runs the Milton Keynes Jam. He is worryingly passionate about switching things on and off.

mrpjevans.com

I f you're blessed with teenagers or others who like to lock themselves away in their room/shed/craft-room, you may well be fed up with shouting up the stairs or out into the garden when you need their attention, especially if headphones are involved. Why not signal your loved ones visually from your phone using a cheap industrial tower light, a Raspberry Pi, and some Python? You'll learn about interfacing Raspberry Pi with normally incompatible devices and how little code is required to create a web server.

This project was inspired by James West's klaxon: **jameswest.site/the-tea-time-klaxon**.

01 Get your Pi ready

This project can be built using any 40-pin version of the Raspberry Pi capable of WiFi and it's perfect for Raspberry Pi Zero W. Once complete, the tower can run 'headless', so no monitor or keyboard will be required. Raspbian Lite is ideal for this as there is no need for a fancy user interface.

Start by installing Raspbian Lite on your microSD card (see **magpi.cc/etcher**), then booting up and running sudo raspi-config. Get WiFi set up and under 'Interfacing', make sure both I2C and SSH are enabled. Finally, set Raspberry Pi's host name under 'Networking' as we'll need it later on.

02 Prepare the Wide Input SHIM

So we can safely power the lights and the Raspberry Pi from a single 12 V PSU, we're using the Pimoroni Wide Input SHIM, a tiny power regulator for Raspberry Pi. To power the lights, we're going 'tap' the SHIM to get a 12 V feed by using the on-board power inputs as a supply. Carefully solder

You'll Need

> 12 V Adafruit 3-Way Tower Light
> **magpi.cc/OUocBH**

> 12 V 2 A PSU with barrel connector
> **magpi.cc/VhOBJY**

> Automation pHAT
> **magpi.cc/RhPiSs**

> Wide Input SHIM (Optional)
> **magpi.cc/RWEDNE**

The light takes its power from the Wide Input SHIM, so only one PSU is needed

The Automation pHAT allows the safe switching of 12 V lights

A close-up of the Wide Input SHIM. Here you can see the positive and negative pads from which we take the 12V current

The Wide Input SHIM is soldered directly to the GPIO pins so the pHAT can sit on top. Note the 12V feed we're taking from the board

two wires to the positive and negative pads on the SHIM. We don't want them shorting out on our Raspberry Pi when the SHIM is in place, so make sure the solder points are trimmed back as much as possible and covered with insulation tape.

03 Install the Wide Input SHIM

The Wide Input SHIM mounts on the first twelve pins of the GPIO. It is intended to sit flush at the base of the pins, so HATs can still be used on top. Therefore, the board needs to be precisely soldered directly on to the GPIO pins.

Place the SHIM onto the GPIO pins, aligning with physical pin 1 (see photo). Secure with masking tape and solder into place starting with pins 1 and 12, then you can remove the tape. Take your time. You should now be able to power up your Raspberry Pi using the barrel connector and 12V supply.

04 Add the Automation pHAT

Although we can now power both Raspberry Pi and the lights from a single 12V supply, Raspberry Pi cannot switch the lights directly without releasing the magic smoke. Pimoroni's Automation pHAT is a general-purpose collection of inputs and outputs for projects such as this one. For our lights, it features three switchable outputs that can handle 12V just fine.

You'll need to solder the supplied GPIO connector and the screw terminals to the pHAT. Check and check again that the orientation is correct. Once ready, attach the pHAT to Raspberry Pi with the Wide Input SHIM sitting in-between.

05 Wire up the tower light

Now our control device is complete, we can attach the tower light itself. There are five wires: one for each colour, a buzzer (that we're not using), and 12V in. Following **Figure 1** (overleaf), connect

the wires to the Automation pHAT. The brown wire is the power in, so we connect that to the 12V feed we took from the SHIM earlier. The three colour wires (red, yellow, and green) are connected to the three screw terminals labelled 'Output Sinking 24V' on the pHAT. Finally, the earth wire from the SHIM needs to connect to the earth on the pHAT.

06 Install Automation pHAT software

Your hardware is ready, so it's time to get some code running. Power up and, from the command line, make sure you're up to date with `sudo apt update && sudo apt upgrade`. Pimoroni takes away all the pain of installing drivers for the Automation pHAT by providing a simple one-line install:

```
curl https://get.pimoroni.com/
automationhat | bash
```

The script takes a few minutes to run. Once completed, allow it to reboot Raspberry Pi (if required) and then you're ready to try things out.

```
cd ~/Pimoroni/automationhat/examples
python output.py
```

Do you see two blinking lights on the tower? If so, it's all working. Press **CTRL+C** to stop.

07 Install Flask

There are many options for remotely controlling the lights, such as Twitter or even email. One of the easiest is to run up a simple web server on your Raspberry Pi itself. To keep the code as simple as possible, we'll use the Flask web application framework for Python, which does all the heavy lifting for us. Installation is as simple as:

```
sudo -H pip install flask
```

If soldered correctly, the SHIM should not interfere with Automation pHAT

Top Tip

Soldering the SHIM

When attaching the SHIM, use a thin tip on your iron and be very careful not to touch components on Raspberry Pi.

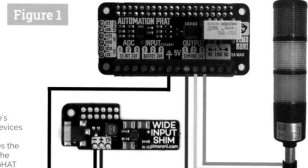

Figure 1

▶ **Figure 1** Here's how all our devices connect. The SHIM provides the power, then the Automation pHAT safely switches the 12 V current

▲ Each line from the light is connected to one of the three 'output' terminals. The ground from the SHIM connects to the ground on the pHAT

Code!

08 From the **/home/pi** directory (if you're not sure, just type cd followed by **RETURN**), issue the following command:

```
mkdir ~/towerlight && cd ~/towerlight
```

Now create a new file:

```
nano towerlight.py
```

Top Tip

Careful with that power!

When using the SHIM, do not attempt to power Raspberry Pi using a regular USB supply at the same time.

Enter the **towerlight.py** code shown here (or you can download it from GitHub). Save the file (**CTRL+X**, then **Y**). When run, this script will fire up a basic web server.

The code uses the Flask and Automation pHAT Python libraries to allow simple switching of the output terminals, allowing the individual lights on the tower to illuminate. Every time a web request comes in, all the outputs are switched off, then the selected one is illuminated.

Testing time!

09 To start the web server, just run:

```
python ~/towerlight/towerlight.py
```

You'll see a few announcements on screen and then it will wait for requests. Using the host name you set earlier, or the IP address of the Raspberry Pi (you can run **ifconfig** to discover this), go to **http://hostname.local:5000/** or **http://ip-address:5000/** from any web browser. If the '.local' address does not work and you're on Windows, you may need to install Apple's Bonjour Services (**magpi.cc/JSUBrT**). The simple website you should now see allows you to switch each light on and off by clicking the links.

Start on boot

10 To start the web server whenever we boot Raspberry Pi, we need to create a service file:

```
sudo nano /lib/systemd/system/towerlight.
service
```

And populate it as follows:

```
[Unit]
Description=TowerLight
After=multi-user.target

[Service]
Type=idle
ExecStart=/usr/bin/python /home/pi/
towerlight/towerlight.py

[Install]
WantedBy=multi-user.target
```

Now, install the service:

```
sudo chmod 644 /lib/systemd/system/
towerlight.service
sudo systemctl daemon-reload
sudo systemctl enable towerlight.service
```

The server will now start automatically on reboot.

Assemble!

11 Once you're happy with testing, you can shorten the leads as you see fit. It's straightforward to mount the tower on top of a small project box and feed the wires through. Mount your Raspberry Pi inside with hot glue or sticky pads, with the large barrel connector lead supplied with the SHIM accessible from the outside.

towerlight.py

> Language: **Python**

```python
001.  import automationhat
002.  import time
003.  from flask import Flask
004.  app = Flask(__name__)
005.
      # Generate a simple HTML page. You could also
006.  use Flash's built-in Jinja Templating.
007.  def makePage(body):
008.      return '''
009.      <!DOCTYPE html>
010.      <html>
011.          <head>
012.              <meta name="viewport"
      content="width=device-width, initial-
      scale=1.0">
013.          </head>
014.          <body>''' + body + '''
015.          <hr/>
016.          <ul>
017.              <li><a href="red">Red</a></li>
018.              <li><a href="yellow">Yellow</a></
019.  li>
                  <li><a href="green">Green</a></li>
020.              <li><a href="off">All Off</a></li>
021.          </body>
022.      </html>
023.      '''
024.
025.  # This function runs every time a request is
      received before routing.
026.  # We switch off all the lights here.
027.  @app.before_request
028.  def allOff():
029.      automationhat.output.one.off()
030.      automationhat.output.two.off()
031.      automationhat.output.three.off()
032.
033.  # Our default home page
034.  @app.route('/')
035.  def index():
036.      return makePage('Index')
037.
038.  # The next three functions switch on the three
      lights
039.  @app.route('/red')
040.  def red():
041.      automationhat.output.one.on()
042.      return makePage('Red')
043.
044.  @app.route('/yellow')
045.  def yellow():
046.      automationhat.output.two.on()
047.      return makePage('Yellow')
048.
049.  @app.route('/green')
050.  def green():
051.      automationhat.output.three.on()
052.      return makePage('Green')
053.
054.  # Switch everything off
055.  @app.route('/off')
056.  def off():
057.      return makePage('Off')
058.
059.  # Let's announce ourselves by making the
      lights blink
060.  automationhat.output.one.on()
061.  time.sleep(0.2)
062.  automationhat.output.one.off()
063.  automationhat.output.two.on()
064.  time.sleep(0.2)
065.  automationhat.output.two.off()
066.  automationhat.output.three.on()
067.  time.sleep(0.2)
068.  automationhat.output.three.off()
069.
070.  # Start the web server on port 5000
071.  if __name__ == '__main__':
072.      app.run(host='0.0.0.0')
```

Congratulations, you now have the perfect teenager/shed-dweller alerting device in a single unit that can be controlled from your phone or any web browser.

Make it your own

12 What we've made here is fairly basic, so there's lots to build on. Why not try improving the website and learning about Flask templates? How about a disco mode? The tower comes with a (very loud) buzzer; could you add it in? What are the other things that could trigger the lights? Twitter? A remote button? Also, don't forget there are plenty of inputs to play with on the Automation pHAT, so the lights could also react to sensors. Check out **magpi.cc/TSEFxL** for an alternative version of the klaxon software with extra features. ◾

GPIO music box

Create a customisable music machine that you control at the touch of a button

MAKER

Marc Scott

Marc is a Senior Learning Manager at the Raspberry Pi Foundation.

raspberrypi.org

In this project, we'll build a button-controlled 'music box' hooked up to a Raspberry Pi's GPIO pins that plays different sounds when different buttons are pressed. Not only does it help teach about using push-buttons and other inputs via GPIO Zero, it's also a good way to learn about playing music or other audio with Python. Let's get started.

01 Set up your project

You will need some sample sounds for this project. There are lots of sound files in Raspbian, but it can be a bit difficult to play them using Python. However, you can convert the sound files to a different file format that you can use in Python more easily.

First, in your home directory (**/home/pi**) create a directory called **gpio-music-box** by right-clicking and selecting New Folder. You will use the new directory to store all your files for the project.

02 Copy the sample sounds

Create a folder called **samples** in your **gpio-music-box** directory.

There are lots of sample sounds stored in the **/usr/share/sonic-pi/samples** directory. In this step, you will copy these sounds into the **gpio-music-box/samples** directory.

Click on the icon in the top-left corner of your screen to open a Terminal window. Type the following command to copy all the files from one directory to the other:

```
cp -r /usr/share/sonic-pi/samples/*
~/gpio-music-box/samples/.
```

When you have done that, you should be able to see all the FLAC (.flac) sound files in the **samples** directory.

03 Convert the sound files

To play the sound files using Python, you need to convert the files from FLAC to WAV format. In a Terminal, move to your **samples** directory:

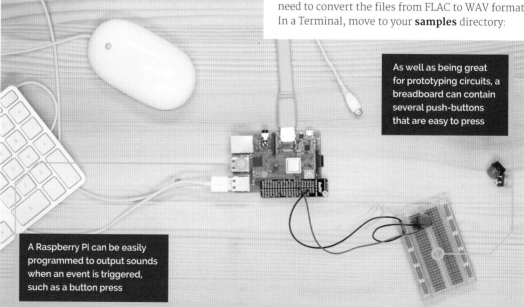

As well as being great for prototyping circuits, a breadboard can contain several push-buttons that are easy to press

A Raspberry Pi can be easily programmed to output sounds when an event is triggered, such as a button press

You'll Need

> Push-buttons

> 220 Ω resistors

> Jumper cables

> Breadboard

> Speakers or headphones

Figure 1

```
cd ~/gpio-music-box/samples
```

Then enter the following commands. This will convert all the FLAC files to the WAV format and then delete the old files.

```
for f in *.flac; do ffmpeg -i "$f"
"${f%.flac}.wav"; done
rm *.flac
```

It will take a minute or two, depending on the Raspberry Pi model that you are using. You should now be able to see all the new .wav files in the **samples** directory.

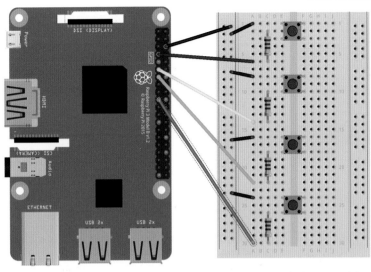

▲ **Figure 1** Here's the way we've wired up our music box

04 Play sounds

Next, you will start to write your Python code. You can use any text editor or IDE to do this — Thonny is always a good choice; you can find it in the Raspbian desktop applications menu (click the top-left raspberry icon) under the Programming category.

To start to create the instruments of your music box, you need to test whether Python can play some of the samples that you have copied. First, import and initialise the Pygame module for playing sound files:

```
import pygame
pygame.init()
```

Save this file in your **gpio-music-box** directory as **musicbox.py**. Choose four sound files that you want to use for your project, for example:

drum_tom_mid_hard.wav
drum_cymbal_hard.wav
drum_snare_hard.wav
drum_cowbell.wav

Then, create a Python object that links to one of these sound files. Give the file its own unique name. For example:

```
drum = pygame.mixer.Sound("/home/pi/gpio-
music-box/samples/drum_tom_mid_hard.wav")
```

Create named objects for your remaining three sounds: cymbal, cowbell, and snare.

Save and run your code. Then, in the shell pane in the Thonny editor, use **.play()** commands to play the sounds. For example:

listing1.py

DOWNLOAD THE FULL CODE: **magpi.cc/musicbox**

> Language: **Python 3**

```
001.  import pygame
002.  from gpiozero import Button
003.
004.  pygame.init()
005.
006.  drum = pygame.mixer.Sound("/home/pi/gpio-music-box/
      samples/drum_tom_mid_hard.wav")
007.  cymbal = pygame.mixer.Sound("/home/pi/gpio-music-box/
      samples/drum_cymbal_hard.wav")
008.  snare = pygame.mixer.Sound("/home/pi/gpio-music-box/
      samples/drum_snare_hard.wav")
009.  bell = pygame.mixer.Sound("/home/pi/gpio-music-box/
      samples/drum_cowbell.wav")
010.
011.  btn_drum = Button(4)
```

```
drum.play()
```

If you don't hear any sound, check that your speakers or headphones are working, properly connected, and that the volume is turned up.

Top Tip

Video tutorial

Check out a shorter, video version of this tutorial on YouTube here: **magpi.cc/ivNfZv**.

05 Connect your buttons

You will need four buttons, wired to GPIO pins on the Raspberry Pi. You'll need to have one side each wired to a different, programmable GPIO pin (not 5V, 3.3V, or GND), and all of them will also need to end at GND, with a resistor somewhere in the circuit.

▶ Try playing your sounds using `.play()` commands in Thonny's shell pane

Place the four buttons into your breadboard. Wire each button to a different numbered GPIO pin. You can choose any GPIO pins you like, but you will need to remember their BCM numbers, which you can find on **pinout.xyz**. Otherwise, just refer to **Figure 1**.

❝ You will need four buttons, wired to GPIO pins on the Raspberry Pi ❞

Top Tip

More music. More buttons

Have more buttons spare and more music you want to use? Create a sample machine or even a fancy keyboard!

06 Play sounds at the press of a button

We're going to use the GPIO Zero Python library to control the buttons. When a specific button is pressed, the program should call a function such as `drum.play()`.

However, when you use an event (such as a button press) to call a function, you don't use brackets `()`. This is because the program must only call the function when the button is pressed, rather than straight away. So, in this case, you just use `drum.play`.

First, set up one of your buttons. Remember to use the numbers for the GPIO pins that you have used – refer to **listing1.py** for how this should look. Remember, you need to import the correct part of GPIO Zero and then define the relevant button. To play the sound when the button is pressed, just add this line of code to the bottom of your file:

```
btn_drum.when_pressed = drum.play
```

Run the program and press the button. If you don't hear the sound playing, then check the wiring of

your button. Now, add code to make the remaining three buttons play their sounds. This should end up looking something like **listing2.py**.

07 Improve your script

The code that you have written should work without any problems. However, it's generally a good idea to make your code a bit cleaner once you have a prototype that works.

The next steps are entirely optional. If you're happy with your script, then just leave it as it is. If you want to make your script a bit cleaner, you can have a go at storing your button objects and sounds in a dictionary, instead of having to create eight different objects.

Have a look at the steps below to learn about creating basic dictionaries and looping over them.

08 Guide to dictionaries

A dictionary is a type of data structure in Python. It contains a series of 'key : value' pairs. Here is a very simple example:

```
band = {'john' : 'rhythm guitar', 'paul'
: 'bass guitar', 'george' : 'lead guitar',
'ringo' : 'bass guitar'}
```

The dictionary has a name, in this case band, and the data in it is surrounded by curly brackets (`{}`). Within the dictionary are the 'key : value' pairs. In this case the keys are the names of the band members; the values are the names of the instruments they play. Keys and values have colons between them (`:`), and each pair is separated by a comma (`,`). You can also write dictionaries so that each 'key : value' pair is written on a new line.

```
band = {
    'john' : 'rhythm guitar',
    'paul' : 'bass guitar',
    'george' : 'lead guitar',
    'ringo' : 'bass guitar'
    }
```

To look up a particular value in a dictionary, you can use its key. So, for instance, if you wanted to find out what instrument ringo plays, you could type:

```
band['ringo']
```

09 Creating a sound dictionary

First, create a dictionary that uses the Buttons as keys and the Sounds as values.

```
button_sounds =
  {Button(4): pygame.mixer.Sound("/home/pi/
gpio-music-box/samples/drum_tom_mid_hard.wav"),
   Button(17): pygame.mixer.Sound("/home/pi/
gpio-music-box/samples/drum_cymbal_hard.wav"),
   Button(27): pygame.mixer.Sound("/home/pi/
gpio-music-box/samples/drum_snare_hard.wav"),
   Button(10): pygame.mixer.Sound("/home/pi/
gpio-music-box/samples/drum_cowbell.wav")}
```

You can now use a `for` loop over the dictionary to tell the program to play the sound when the button is pressed:

```
for button, sound in button_sounds.items():
    button.when_pressed = sound.play
```

The final code can be found in **listing3.py**. Have fun with your new musical Raspberry Pi! M

▲ There are many Python IDEs and text editors, but Thonny is easy to use and already in Raspbian

listing2.py

> Language: **Python 3**

```
001.  import pygame
002.  from gpiozero import Button
003.
004.  pygame.init()
005.
006.  drum = pygame.mixer.Sound("/home/pi/gpio-music-box/
      samples/drum_tom_mid_hard.wav")
007.  cymbal = pygame.mixer.Sound("/home/pi/gpio-music-box/
      samples/drum_cymbal_hard.wav")
008.  snare = pygame.mixer.Sound("/home/pi/gpio-music-box/
      samples/drum_snare_hard.wav")
009.  bell = pygame.mixer.Sound("/home/pi/gpio-music-box/
      samples/drum_cowbell.wav")
010.
011.  btn_drum = Button(4)
012.  btn_cymbal = Button(17)
013.  btn_snare= Button(27)
014.  btn_bell = Button(10)
015.
016.  btn_drum.when_pressed = drum.play
017.  btn_cymbal.when_pressed = cymbal.play
018.  btn_snare.when_pressed = snare.play
019.  btn_bell.when_pressed = bell.play
```

listing3.py

> Language: **Python 3**

```
001.  import pygame
002.  from gpiozero import Button
003.
004.  pygame.init()
005.
006.  button_sounds = {Button(4): pygame.mixer.Sound("/home/pi/
      gpio-music-box/samples/drum_tom_mid_hard.wav"),
007.              Button(17): pygame.mixer.Sound("/home/
      pi/gpio-music-box/samples/drum_cymbal_hard.wav"),
008.              Button(27): pygame.mixer.Sound("/home/
      pi/gpio-music-box/samples/drum_snare_hard.wav"),
009.              Button(10): pygame.mixer.Sound("/home/
      pi/gpio-music-box/samples/drum_cowbell.wav")}
010.
011.  for button, sound in button_sounds.items():
012.      button.when_pressed = sound.play
```

Laundry-saving
Rain Detector

Save your washing from a soaking. This easy-to-build
wire-free rain detector alerts you to sudden downpours

MAKER

PJ Evans

PJ is a software engineer and tinkerer who has littered his house with Raspberry Pi devices. He mostly has nice dry clothes.

mrpjevans.com

There's nothing quite like clean air-dried clothes fresh from the line, unless an unexpected shower ruins everything. Ever heard that cry of "RAIN!" from a member of the household, only to be followed by the thundering of feet down the stairs in a desperate bid to save your Sunday best from another trip to the washing machine?

Catch the rain as soon as it starts with this simple standalone build that alerts your phone as soon as it detects raindrops. There's no soldering required, just a few cables. We need low power consumption and WiFi, so this is a perfect project for a Raspberry Pi Zero W.

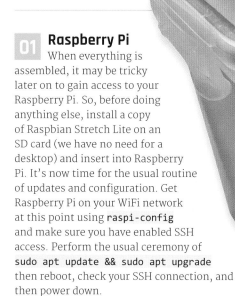

Raspberry Pi, controller, and power are kept safe in the airtight box

The rain sensor works by water shorting the connection. Two are used to increase surface area

You'll Need

> 2 × Rain sensor boards with one controller **magpi.cc/ pMUaWu**

> Small breadboard

> Small USB power bank e.g. **magpi.cc/ iYvwEL**

> Airtight small food container

> Jumper cables

01 **Raspberry Pi**

When everything is assembled, it may be tricky later on to gain access to your Raspberry Pi. So, before doing anything else, install a copy of Raspbian Stretch Lite on an SD card (we have no need for a desktop) and insert into Raspberry Pi. It's now time for the usual routine of updates and configuration. Get Raspberry Pi on your WiFi network at this point using `raspi-config` and make sure you have enabled SSH access. Perform the usual ceremony of `sudo apt update && sudo apt upgrade` then reboot, check your SSH connection, and then power down.

Top Tip 👍

Keep Raspberry Pi dry

A critical part of this project is ensuring that, in the event of a downpour, your Raspberry Pi isn't ruined by rain. Make sure the hole for the wires is small and covered.

▲ Here's everything you need to build your rain detector. Try to use a container with a rubber or silicone seal

02 Mount the sensors to the lid

You can use any number of sensors you wish, but two works well. Either secure the two plates to the lid using insulation or duct tape. Alternatively, 3D-print the enclosure pictured (STL files available from **magpi.cc/DAuqUT**) and secure with glue or sticky pads.

Two pairs of jumper cables need to be attached; one to each sensor plate. Polarity does not matter. The other end of the cables will have to thread into the container, so make as small a hole as possible in an appropriate place so the wires can get through, minimising the chance of any water ingress.

03 Connect the sensors to the controller

In order for Raspberry Pi to understand what's going on, a small controller board (supplied with the sensors) is required. This takes the small current that is shorted by water and converts it into a digital signal. Using the breadboard, connect the two pairs of wires from the sensors in parallel (so that either sensor could make the circuit) and then insert the controller's receiving pins (the side with two connectors) into the breadboard so that each pin connects with one wire from each sensor.

04 Connecting the controller

To complete our circuit, take a careful look at the four pins on the controller board – marked as Ao, Do, GND, and VCC. Using some jumper wires, hook the controller up to Raspberry Pi as follows: VCC to GPIO pin 2 (5 V), GND to any GND on the GPIO (e.g. pin 6) and Do to GPIO 17 (pin 11). Do and Ao are two different ways of reading output from the sensor. Do is a straight digital on or off, the threshold being controlled by the variable resistor on the board. Ao is an analogue output that (when converted to digital) ranges between 0 and 1024 depending on how heavy the rain is.

▼ The sensor plates need to be connected to the controller board in parallel, not in series. That way, either one can trigger the alert. Polarity is unimportant

Top Tip

05 Assembly

Connect the micro USB cable from your power bank to the power input on Raspberry Pi and arrange everything inside your container. Ideally things shouldn't move about, so keep everything in place with sticky pads or tack. You should now be able to seal the container with everything inside, the wires to the sensor plates coming out without being squashed or stressed. Once you're happy, open it up and attach the power bank, then close it again and check your connection. The power bank, depending on its rating, should keep your Raspberry Pi Zero W alive for a few hours at least.

06 Software

Add the script printed here and save it as **rainbot.py** (or download from GitHub) into a convenient spot such as **~/pi/rainbot**. Once in place, perform an initial test by running `python3 ~/pi/rainbot/rainbot.py`. You should see a readout every five seconds: 'True' if it's dry, 'False' if it's wet. Press **CTRL+C** to stop the script.

07 Pushover

To get alerts, we're going to use Pushover, a neat one-time-payment notification service for smartphones (there's a seven-day free trial). Sign up at **pushover.net**. When logged in, you'll see a 'User Key'; make a copy of this. Now follow the instructions to create an 'Application Token'. You'll be given an API key. Edit the script to replace the API key values, where prompted, with the keys you have been given. Make sure the Pushover app is installed on your phone.

Run the script again. This time, wet one of the panels slightly. A light should illuminate on the controller. If all is well, a few seconds later your phone will display an alert.

08 Run automatically

Let's set the script to run on startup. Create the following file as a superuser:

```
sudo nano /lib/systemd/system/rainbot.
service
```

▶ The controller takes the output from the two panels and converts it into a digital signal. This is then connected to Raspberry Pi, which also supplies 5 V to the controller

▲ Make sure everything fits well, is secured and that the jumper cables are not going to pop out

Add in the following text:

```
[Unit]
Description=Rainbot
After=multi-user.target

[Service]
Type=idle
ExecStart=/usr/bin/python3 /home/pi/
rainbot/rainbot.py

[Install]
WantedBy=multi-user.target
```

Press **CTRL+X** to save and quit out of nano. Now issue the following commands:

```
sudo chmod 644 /lib/systemd/system/
rainbot.service
sudo systemctl enable rainbot.service
sudo systemctl daemon-reload
```

Reboot Raspberry Pi. The script will start on reboot (although you won't see any output). Test it with water again.

09 Make it your own

There are lots of improvements that can be made, which we'll leave up to you to explore. Pushover is convenient, but the function could be easily replaced with, well, anything you like. The frequency of checks could be altered (it's currently every five seconds). How about adding an analogue to digital converter and use the Ao output to gauge how heavily it's raining? It's also a great start for a weather station project if you start recording the data. One useful addition would be adding a button for safe shutdown of Raspberry Pi after use. M

rainbot.py

> Language: **Python 3**

DOWNLOAD THE FULL CODE:
magpi.cc/DAuqUT

```python
001. from gpiozero import DigitalInputDevice
002. from time import sleep
003. import http.client, urllib.parse
004.
005. # Some setup first:
006. APP_TOKEN = 'YOUR_PUSHOVER_APP_TOKEN'      # The app token -
      required for Pushover
007. USER_TOKEN = 'YOUR_PUSHOVER_USER_TOKEN'    # Ths user token -
      required for Pushover
008.
009. # Set up our digital input and assume it's not currently raining
010. rainSensor = DigitalInputDevice(17)
011. dryLastCheck = True
012.
013. # Send the pushover alert
014. def pushover(message):
015.     print(message)
016.     conn = http.client.HTTPSConnection("api.pushover.net:443")
017.     conn.request("POST", "/1/messages.json",
018.       urllib.parse.urlencode({
019.         "token": APP_TOKEN,     # Insert app token here
020.         "user": USER_TOKEN,     # Insert user token here
021.         "title": "Rain Detector",
022.         "message": message,
023.       }), { "Content-type": "application/x-www-form-urlencoded" })
024.     conn.getresponse()
025.
026. # Loop forever
027. while True:
028.
029.     # Get the current reading
030.     dryNow = rainSensor.value
031.     print("Sensor says: " + str(dryNow))
032.
033.     if dryLastCheck and not dryNow:
034.
035.         pushover("It's Raining!")
036.
037.     elif not dryLastCheck and dryNow:
038.
039.         pushover("Yay, no more rain!")
040.
041.     # Remember what the reading was for next check
042.     dryLastCheck = dryNow
043.
044.     # Wait a bit
045.     sleep(5)
```

Smart door

Adding a Raspberry Pi to your door has magical results. Want to see who's at the door or know when the post has arrived? Control the lock? Read on...

PJ Evans

PJ is a writer, software engineer, and Raspberry Jam organiser. Too much of his house now thinks for itself.

mrpjevans.com

Is your door a bore? Open and close, open and close. Snoozefest. Surely it can do more than that? How about a smart door that knows when someone approaches, when the post arrives, and can even offer remote viewing of the peephole? You can also add intelligent lighting, a controllable door lock, and facial recognition, all powered with your Raspberry Pi. So, let's ignore super-expensive door systems and build our own. You can do as much, or as little, as you like of this project and there's plenty of room for new and inventive uses.

01 Prepare your Raspberry Pi

Although you can use any WiFi-capable Pi, this is a perfect project for the new Raspberry Pi 3A+. Start by attaching Raspberry Pi to the Touch Display and preparing a microSD card with the latest Raspbian Stretch release. To allow easier access and mounting, we've detached the control board from the back of the screen, taking great care of the ribbon cable. Eventually, they'll be put in a smart 3D-printed case. Now, get your Pi set up and make sure to `sudo apt update && sudo apt upgrade` before proceeding.

02 Attach the camera

We're going to keep an eye on the outside world by replacing the door's peephole with the Raspberry Pi camera. A peep-hole is typically a two-piece barrel that screws together and can be easily unscrewed from the inside. Remove the barrel and cover the hole with the camera. We're just going to affix this with tape for now; a printed mount will come later. Mount the screen and Pi to the door (we used 3M Command strips), placed so you can attach the camera's ribbon cable to Raspberry Pi once it's shut down. Make sure the camera is enabled in Raspberry Pi Configuration or raspi-config.

03 Footsteps approaching!

The first smart thing our door is going to do is detect someone approaching it. A cheap PIR sensor is perfect for the job. These cool little geodesic domes are triggered by heat and are the same gizmos that you find in motion-sensor lights, switches, and security systems. Connect to Raspberry Pi as in **Figure 1**, checking whether you have a 5V or 3.3V sensor. Sensitivity and duration of a 'detection' can be controlled by the two potentiometers on the PIR board. Mount this outside in a suitable location to 'watch' your door.

You'll Need

- Raspberry Pi Touch Display **magpi.cc/touch**
- Camera Module **magpi.cc/camera**
- PIR sensor **magpi.cc/vmqYLG**
- 2 × Security door contact reed switch **magpi.cc/FDjbna**
- Wired doorbell **magpi.cc/KFYWcQ**
- PAM8302 amplifier **magpi.cc/mifFLc**
- Speaker **magpi.cc/PwkasX**
- Magnetic access control system **magpi.cc/rkEXYF**

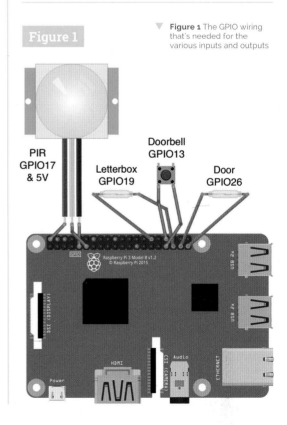

Figure 1 The GPIO wiring that's needed for the various inputs and outputs

04 Monitor the door and letterbox

We have two magnetic reed switches, the type you find on windows and doors for security systems. They are made up of two parts: the wired part is a reed switch and the unwired a magnet. When the magnet meets the switch, it closes. If we attach the magnet to the door and the switch to the frame, when the door opens, so does the switch. There's no polarity to worry about, so connect one wire to GPIO 26 and the other to the adjacent ground. Repeat for the letterbox using GPIO 19. You may need a breadboard.

05 Ding dong!

Regular doorbells? Yawn. If we replace the doorbell with our own button, we can take a photo with Raspberry Pi Camera Module when someone presses it and send a notification. Way better. Mount a standard wired doorbell, which after all is just a momentary contact button, to the outside door frame and wire it back to Raspberry Pi using GPIO 13 and a GND pin. If you're prototyping on a breadboard, a tactile switch will do fine.

06 Sounds good

There's little point in a doorbell that makes no sound. We can use the small, but surprisingly powerful, PAM8302 amplifier with a speaker to make some noise. Supply power by soldering 'Vin' to an available 3V3 pin on Raspberry Pi, and ground to GND. To get an audio signal, you can tap the audio connector's signal and ground, then connect them to A+ and A- respectively. Finally, solder the speaker to the larger + and - terminals. When prototyping, you can skip this and use any active or passive speaker via the audio connector on Raspberry Pi.

07 Code

Double-check all your connections and power on Raspberry Pi. To use the code published here (overleaf), open a Terminal and enter:

```
mkdir ~/smartdoor
nano ~/smartdoor/smartdoor_test.py
```

Now type in the code as shown. Alternatively, to download all the code:

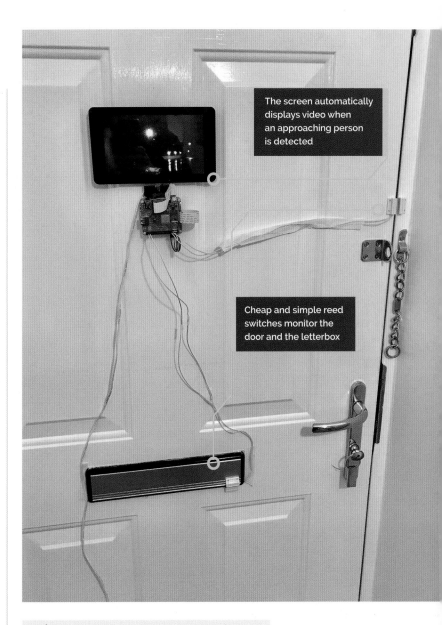

The screen automatically displays video when an approaching person is detected

Cheap and simple reed switches monitor the door and the letterbox

```
cd
git clone https://github.com/mrpjevans/
smartdoor
```

To enable it to play our doorbell sample:

```
sudo apt install mpg123
```

Now test with:

```
python3 ~/smartdoor/smartdoor_test.py
```

Watch the console output. If everything is working, you should be able to trigger the PIR, the reed switches, and the doorbell. The camera will capture ten seconds of video when motion is detected, and a photo when the doorbell is pressed. These are both saved to the desktop.

Top Tip

Night is dark

If you want the camera to work well at night, you may want to consider a Pi NoIR Camera Module supported with some infrared lighting.

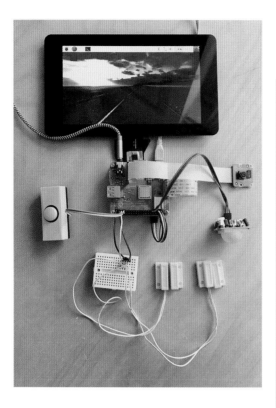

▶ You may want to prototype this project and test it before taking a drill to your door!

⊼ The web app can run on the touchscreen, as well as on mobile devices or desktop browsers. Release the door from anywhere!

10 Door lock

If you're interested in being able to control your door's lock, you may see that some solutions are very pricey. One that is perfect for experimentation is the magnetic hold lock, which uses an electromagnet to hold the door closed. The one we've used can withstand 180 kg of force, although stronger ones are available. The magnet mounts on the door and the electromagnet on the frame. The provided PSU contains a relay that can be powered by Raspberry Pi by simply connecting it to a spare GPIO line and ground. Please note this is no replacement for a proper door lock system.

11 Web app

If would be great to see what our door has been up to remotely, so a web app seems the next logical step. In the directory called **webapp** is a Python script that uses Flask to provide a web server that is usable on mobile devices. You can take a photo from the peephole, see the last recorded video, and even control the magnetic door lock from Step 10. Simply run the app alongside the others. Better still, set **smartlights.py**, **porch.py**, and **webapp/smartdoor.py** to start on boot (see the repository README).

08 Get alerts!

Let's make this useful. Install Pushover on your phone, head over to **pushover.net**, sign up for a trial account, then log in and make a note of your User Key (a long string of characters). Now create a new Application and give it a name. Once created, you'll see an API Token. Make a note of this too. From the GitHub repository, edit **smartdoor.py** and add the User Key and API Token where shown. Run this version and you'll get phone alerts for each event and even a photo attachment when the doorbell is pressed.

09 Intelligent porch light

Following on from the Trådfri lighting tutorial in *The MagPi* #75 (**magpi.cc/75**), if you have an external porch light, why not make it smart! The file **porch.py** will connect a Trådfri smart light to an API that provides sunrise and sunset times for your location. Leave the script running and the light will switch on and off at the correct times. Additionally, it monitors the PIR sensor and will switch to full brightness when someone approaches! To use the script, get your latitude and longitude (you can use Google Maps or Earth) and edit **porch.py** as directed in the file.

Top Tip 👍

Get the right lock

Magnetic door locks vary in size and shape; measure twice and order once!

smartdoor_test.py

> Language: **Python 3**

```python
001.  from picamera import PiCamera
002.  from gpiozero import MotionSensor
003.  from gpiozero import Button
004.  from time import sleep
005.  import os
006.  import subprocess
007.  import sys
008.
009.  print('Getting smart...')
010.
011.  # Set up all our devices
012.  camera = PiCamera()
013.  motion = MotionSensor(17)
014.  doorSensor = Button(26)
015.  letterbox = Button(19)
016.  doorbell = Button(13)
017.
018.  def motionDetected():
019.      print('Motion detected, video recording')
020.      os.system('DISPLAY=:0 xset s reset')  # Wakes
      the display up
021.      camera.start_preview()
022.      camera.start_recording(
      '/home/pi/Desktop/motion.h264')
023.      sleep(10)
024.
025.  def motionStopped():
026.      print('Stopping video recording')
027.      camera.stop_recording()
028.      camera.stop_preview()
029.
030.  def doorOpen():
031.      print('Door open')
032.
033.  def doorClosed():
034.      print('Door closed')
035.
036.  def letterboxOpen():
037.      print('You got mail!')
038.
039.  def doorbellPressed():
040.      subprocess.Popen(['mpg123', '/home/pi/
      smartdoor/doorbell.mp3'],
041.                       stdout=subprocess.PIPE,
      stderr=subprocess.STDOUT)
042.      camera.capture('/home/pi/Desktop/doorbell.jpg')
043.      print('Someone\'s at the door!')
044.
045.  # Attach our functions to GPIOZero events
046.  motion.when_motion = motionDetected
047.  motion.when_no_motion = motionStopped
048.  doorSensor.when_pressed = doorClosed
049.  doorSensor.when_released = doorOpen
050.  letterbox.when_released = letterboxOpen
051.  doorbell.when_released = doorbellPressed
052.
053.  print('Smart door is smart')
054.
055.  # Loop forever allowing events to do their thing
056.  try:
057.      while True:
058.          pass
059.  except KeyboardInterrupt:
060.      print('Smart door no longer smart')
061.  except:
062.      print('Oh dear')
```

12 Facial recognition

Once a futuristic technology, decent facial recognition is now well within the grasp of Raspberry Pi. Using the doorbell photo taken by Raspberry Pi, we can recognise a face using reference photos and send an alert to Pushover with the name of the caller! In a secure environment, a recognised face could even trigger the lock or you could play a welcome announcement. The install process is a little complicated, so if this interests you, see the documentation in the GitHub repository in the **face_recognition** directory of the 'smartdoor' repository.

13 Over to you

Here we've given you the basics to get going, but more complex events are possible. You could alert different people based on facial recognition or play custom doorbell tones. And, if you had problems with deliveries, video evidence can build up automatically. On a serious note, remember a lot of this is 'just for fun' and designed to inspire, so unless you're prepared to put in the work hardening the code and including failsafes, don't rely on this, or possibly make it as a fun kids' door project (but maybe without the lock!).

NeoPixel display
cabinet lights

Light up a display cabinet with some NeoPixels, a Raspberry
Pi, and a little Python code

MAKER

**Rob
Zwetsloot**

Rob is amazing.
He's also the
Features Editor
of *The MagPi*, a
hobbyist maker,
cosplayer, comic
book writer, and
extremely modest.

magpi.cc

You'll Need

> NeoPixel lights

> Circuit wire

> Push-button

> 470 Ω resistor

> 5 V power source

> Soldering iron

We've covered NeoPixels in *The MagPi*
before, with some wonderful cosplay
lights (**magpi.cc/45**) and Christmas
tree lights (**magpi.cc/52**), which we've used
ourselves. It's been a long time since we've
controlled any NeoPixels with a Raspberry Pi,
though – so long that there's actually a newer and
much easier method for doing it.

We thought it was high time to try this out, and
get a display cabinet lit up fancily in the process.

to create a little semicircle in our cabinet for a bit
more interesting coverage.

If you have a big cabinet and wish to line the
entire thing, you can always get a long flexible
strip of NeoPixels. You can even get single lights,
or smaller circles of NeoPixels.

The only thing you need to make sure you do
for all the NeoPixel types is correctly count the
number of LEDs in your system. We'll tell you why
in a few steps' time.

01 Choose your NeoPixels

There are many configurations and types
of NeoPixels that you can buy. For our display
cabinet, we chose two quarter-circle stripes of
lights that hold 15 LEDs a piece. This allowed us

02 Choose a location

What do you want to light up? For our
project, a shelf of figurines was all we wanted to
illuminate, and so we decided to add the lights
above the shelf – attached to the 'ceiling', as
it were.

You'll need to take into consideration light
coverage and camouflage in your display cabinet.
Think about sight lines if you want to keep them
hidden from specific angles, and look to see if your
display cabinet has anything to aid in adding lights
– the IKEA Detolf has a little plug on the top for
wires, for example.

There also needs to be access for Raspberry Pi to
control the lights, so keep that in mind.

03 Assemble your circuit

We've put together a handy circuit diagram
(**Figure 1**) for you to follow along to. There are a few
important things to note, though, to make sure you
understand it.

The NeoPixel strips have three pads on them:
one for 5 V power, one for ground, and a 'data'
port. The data needs to be connected to the GPIO

Figure 1

▲ **Figure 1** This diagram shows
the rough circuit for our
setup, and should be used as
a guide for your own

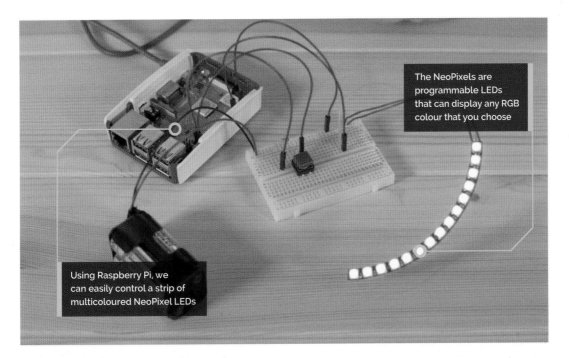

The NeoPixels are programmable LEDs that can display any RGB colour that you choose

Using Raspberry Pi, we can easily control a strip of multicoloured NeoPixel LEDs

Top Tip

RGB colour

Red, green, and blue lights make up a single NeoPixel. Using a value of each colour between 0 and 255, you can create an entire rainbow of colours.

pin we're sending the signals from. You also need to make sure to connect it to the Din (data in) pad. If you're chaining together strips like we've done, make sure you connect the Dout (data out) from the first strip to the Din of the next strip. It's also good practice for ground on the NeoPixels to go to Raspberry Pi's ground, as well as the ground of the power source.

We've placed the button way down the GPIO pins to keep it clear. You need the resistor in the little button circuit to make sure it works properly and can be sensed by Raspberry Pi.

04 Soldering the NeoPixels

This bit can be tricky, but you will have to solder some wires to your NeoPixel strips. Make sure your soldering iron is properly prepared if this is your first time using it (check this video: **magpi.cc/GCUNyL**). Also, if this is your first time soldering, check out the Raspberry Pi Foundation's video on it: **magpi.cc/Ahvxdk**.

We recommend putting a little solder on your wire (this is called tinning it), as well as a dab on the pads. That way, you just need to heat up the solder already on the wire and pad so that they fuse and connect.

05 Powering the system

You need to power two parts of your light display: Raspberry Pi and the NeoPixels. The

" You need to power two parts of your light display: Raspberry Pi and the NeoPixels "

NeoPixels require a 5V input, and also draw a lot of power at the same time; so, to be safe, you really shouldn't power more than two or three NeoPixels with the 5V pin of Raspberry Pi. In our circuit, we've used four 1.5V AA rechargeable batteries, as they're technically closer to about 1.2V each – that means we get roughly 5V out of them.

A more convenient way is to get a 5V power supply and use a power terminal with two screw ports to attach the positive and negative ends. Please make sure the power supply is unplugged while you do this, and be very careful when plugging it in.

You can technically run a Raspberry Pi using the GPIO pins; however, we have elected to run a micro USB cable up to our Raspberry Pi.

06 Basic code

We've written the code, **rollcall.py**, to work with our specific system. Download it from **magpi.cc/DisplayLights** and we'll run through it.

First, we import the generic stuff: `time`, `gpiozero`, as well as the `neopixel` library and its related `board` library. We've also used the `numpy` library so that we can create RGB values for the LEDs, which enables them to fade between colours.

▶ We used Blu Tack to attach the lights to the top of our cabinet

Our system uses 30 LEDs, and we've connected it to GPIO 18 on the board. We've then defined six colours that represent the figurines in the display for our added colour cycle effect. After that, we tell the code about the strip of NeoPixels.

We then define how to calculate the values for colour fading, and the colours to transition between, before turning the strip of LEDs to pure white. Finally, we set the main `rollcall_cycle` function to loop, so we can press the button whenever we want.

Top Tip

Roll call

The six colours chosen represent the main team of figurines in the cabinet – it's a tradition in Japanese superhero TV shows to run down your name and colour in a specific order.

07 Alter your code

The main parts of the code to pay attention to are the `LED_COUNT`, `LED_PIN`, and `button` values. Your number of LEDs will likely differ from ours, and you may have connected the strip and button to different GPIO pins.

As for the colour cycle, play around with that to your heart's content, or remove it completely! You can even change the main colour value for the standard lights, using RGB values from 0 to 255.

08 Testing your lights

Before attaching everything to your cabinet, we highly recommend testing your LEDs. Run the code from within your preferred Python IDE, and make sure that not only are you getting the correct

colours (you may have a GRB set of NeoPixels instead of RGB, for example), but that also the button works.

09 Final Raspberry Pi prep

Our setup's Raspberry Pi is going to be headless, which means we want the Python code to load up after Raspberry Pi turns on. Our preferred method is to add a line to **/etc/profile** – it makes it a lot easier. Open up a Terminal window and type:

```
sudo nano /etc/profile
```

Use the arrow key to go to the bottom, then add:

```
sudo python rollcall.py
```

If you've saved the Python script to a specific folder other than the home directory, make sure to include the path to it as well. Save and exit the file. You can also turn off 'boot to desktop' in Raspberry Pi configuration settings, so that the entire system loads up faster.

10 Mounting the lights

Depending on your NeoPixels, you can attach them in several different ways. We've used Blu Tack to stick ours to the ceiling of our display

rollcall.py

> Language: **Python**

```python
001.  #!/usr/bin/env python
002.
003.  import time
004.
005.  from gpiozero import Button
006.
007.  import board
008.  import neopixel
009.  import numpy as np
010.
011.  button = Button(21)
012.
013.  # LED strip configuration:
014.  LED_COUNT   = 30      # Number of LED pixels.
015.  LED_PIN     = board.D18     # GPIO pin
016.  LED_BRIGHTNESS = 0.2 # LED brightness
017.  LED_ORDER = neopixel.GRB # order of LED colours.
      May also be RGB, GRBW, or RGBW
018.
019.  # The colour selection selected for this
      project: red, blue, yellow, green, pink, and
      silver respectively
020.
021.  gokai_colours = [(255,0,0),(0,0,255),(255,255,0)
      ,(0,255,0),(255,105,180),(192,192,192)]
022.
023.  # Create NeoPixel object with appropriate
      configuration.
024.  strip = neopixel.NeoPixel(LED_PIN, LED_COUNT,
      brightness = LED_BRIGHTNESS, auto_write=False,
      pixel_order = LED_ORDER)
025.
026.  # Create a way to fade/transition between
      colours using numpy arrays
027.
028.  def fade(colour1, colour2, percent):
029.      colour1 = np.array(colour1)
030.      colour2 = np.array(colour2)
031.      vector = colour2-colour1
032.      newcolour = (int((colour1 + vector*percent)
      [0]), int((colour1 + vector * percent)[1]),
      int((colour1 + vector * percent)[2]))
033.      return newcolour
034.
035.  # Create a function that will cycle through the
      colours selected above
036.
037.  def rollcall_cycle(wait):
038.      for j in range(len(gokai_colours)):
039.          for i in range(10):
040.              color1 = gokai_colours[j]
041.              if j == 5:
042.                  color2 = (255,255,255)
043.              else:
044.                  color2 = gokai_colours[(j+1)]
045.              percent = i*0.1   # 0.1*100 so 10%
      increments between colours
046.              strip.fill((fade(colour1,colour2,
      percent)))
047.              strip.show()
048.              time.sleep(wait)
049.
050.  strip.fill((255,255,255))
051.  strip.show()
052.
053.  # Main function loop
054.
055.  while True:
056.
057.      time.sleep(1)
058.
059.      button.wait_for_press()
060.      rollcall_cycle(0.2)     # 0.2 seconds between
      colour updates
```

case; however, you could also use glue. For the long strips, you can always nail them in with staples so that you don't actually have to go through the strip.

Just make sure any exposed PCB is not touching anything conductive.

11 Permanent circuit tips

You don't want exposed wires everywhere just to light up your display case. Creating a 3D case to house Raspberry Pi is a good first step, and using heat-shrink tubing to encase all the cables makes the whole thing a lot neater. You can also cover any soldered joints with a bit of hot glue.

12 Have fun and experiment!

This basic setup can very easily be expanded upon. As Raspberry Pi is internet-connected, you can use some IoT stuff like Twitter triggers or noise activation or temperature-dependent colouration. You can even add more strips to Raspberry Pi so that you can have several layers of lighting effects. We hope this really improves your display cabinets. 🗚

Build a secret
radio chat device

Add a cheap 433MHz radio to your Raspberry Pi to send wireless messages without WiFi and operate remote-control main sockets

MAKER

PJ Evans

PJ is a writer, developer, and runs the Milton Keynes Jam. He is worryingly passionate about switching things on and off.

mrpjevans.com

WiFi is all well and good, but is it the only option for wireless communication on Raspberry Pi? What if there isn't a network available or you need a longer range? 433MHz radio is where you want to be. In this tutorial we'll add this capability to a pair of Raspberry Pi boards and show how to send wireless messages from one to the other with no WiFi network. Then, we'll increase the range with a touch of science and start talking to RF-based switchable main sockets. Have Raspberry Pi-controlled sockets all around the house!

You'll Need

- 2 × 433MHz transceivers
  magpi.cc/yXnbtu

- 2 × Mini breadboards
  magpi.cc/QEyLck

- 12 × M/F jumper leads
  magpi.cc/UgGBxg

- RF mains socket kit
  magpi.cc/vuCjwL

▶ Here's everything you need. It is possible to just use one Raspberry Pi, but double up for more fun

01 Prepare Raspberry Pi boards

To demonstrate sending messages using 433MHz, it makes sense to use two Raspberry Pi boards so we can have a conversation. None of what we're doing here requires much processing power, so any Raspberry Pi will do, even original Model As or Bs. Depending on what you're comfortable with, install either full Raspbian or – as we're doing here – Raspbian Lite, as

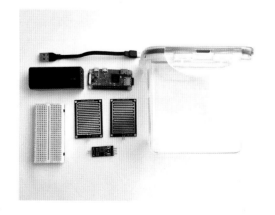

everything will be run from the command line. If you haven't got access to multiple monitors and keyboards, consider using SSH to access each Raspberry Pi with two windows on your main computer. That way you can see everything taking place.

02 Meet the transceivers

Each kit comes with two circuit boards. The longer of the two boards is the receiver, sporting four pins. Before wiring up, check the labelling of these pins very carefully as they do sometimes vary. Regardless of position, there will be 5 V power in (labelled VCC), ground (GND), and two 'DATA' lines which transmit the received signals. These are identical so you can use either.

The smaller transmitter has three lines, which again can vary in position based on the manufacturer. Just like the receiver, you have VCC for power, GND for ground, and this time, a single data line.

03 Wire-up the breadboard

We're using a tiny breadboard, but any size will work. In fact, a larger board with power and ground rails might be a bit tidier. Carefully insert a receiver and transmitter in each breadboard alongside each other. We want the two breadboards opposite so that the transmitter of Raspberry Pi #1 (which we're calling 'Alice') is pointing directly at the receiver of Raspberry Pi #2 ('Bob') and vice versa.

Connect six jumper leads to each breadboard, one on the rail for each pin of the transceiver pair. It doesn't matter which 'DATA' line you use on the receiver.

Initially, the two Raspberry Pi boards must be really close to communicate. We'll increase the range later

The 433MHz receiver and transceiver come as a pair of small PCBs

Warning!
LPD433 radio

Low power device 433MHz radio may require a licence in some countries. For more details, visit **magpi.cc/H4sHcF**

04 Connect to Raspberry Pi boards

Connect each Raspberry Pi to its six jumper leads. Both the receiver and transmitter run at 5 V, so connect each VCC jumper lead to physical pins 2 and 4 of the GPIO (the top two right-hand pins when pin 1 is top-left). Next, connect the GND leads to pins 6 and 9. Although your radio is now powered, it's not much use if it can't send and receive data, so connect the transmitter's DATA to GPIO 17 and the receiver's DATA to GPIO 27 (pins 11 and 13).

05 Test receive

Before we can do anything with our newly installed radio, we need some software. Open up a Terminal and issue the following commands:

```
cd
sudo apt install python3-pip git
pip3 install rpi-rf
git clone https://github.com/mrpjevans/
rfchat.git
```

You now have everything installed to test your hardware. Pick your favourite of the two Raspberry Pi boards and enter the following:

```
cd ~/rfchat
python3 receive.py
```

Hold the remote from the RF kit close to the receiver and press its buttons. See numbers appear? Great. If not, review your wiring. Press **CTRL+C** to quit and repeat on the other Raspberry Pi.

06 Test send

Position the Raspberry Pi boards so the two breadboards are within a centimetre of each other, with Alice's transmitter pointing at Bob's receiver and likewise the other way around. On Alice, start the receive script just as we did in the previous step. On Bob, enter the following in the Terminal:

```
cd ~/rfchat
python3 send.py 1234
```

All being well, '1234' should be displayed repeatedly on Alice's screen. There's no error correction, so it's normal to see missing or corrupt characters. If it doesn't look quite right, try again. Once you're happy, reverse the test to confirm Bob's receiver is also working.

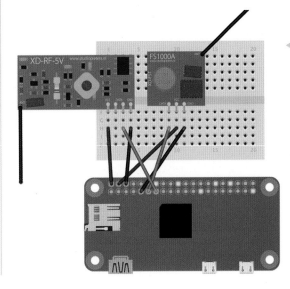

◀ The pair of transceivers do not require any additional components and can be wired straight to the GPIO

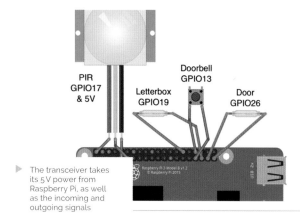

PIR
GPIO17
& 5V

Doorbell
GPIO13

Letterbox
GPIO19

Door
GPIO26

The transceiver takes its 5V power from Raspberry Pi, as well as the incoming and outgoing signals

07 Let's have a chat

Our two Raspberry Pi boards can now communicate wirelessly without WiFi. To demonstrate what's possible, take a look at the **rfchat.py** script. This code uses threading (code-speak for doing multiple things at once) to monitor the keyboard and receiver for data. We convert incoming and outgoing data to numbers (ASCII) and back. The result is a live chat interface. You can now send and receive messages. To start:

```
cd ~/rfchat
python3 rfchat.py
```

Now type slowly on either Raspberry Pi and

the message will appear on the other. In fact, your local output is your receiver picking up your own transmitter!

08 Increasing range with science

The reason for the radio's poor range is the tiny antennas, but this can be fixed. The antenna's length needs to be a harmonic of the wavelength, which is calculated by dividing the speed of light by the frequency (299 792 458 m/s divided by 433 000 000). You can keep dividing the result of 692.36 mm by 2 until you get a sensible length. A 173 mm antenna is long enough to give an impressive range, normally covering a whole house. Solder 173 mm wires to all four 'ANT' solder points on the PCBs. Your rfchat should now work over long distances.

09 Socket to me

There are many household devices that use 433MHz to send control codes. Among the most popular are remote-control mains sockets,

rfchat.py

> Language: **Python 3**

```python
001.  import sys
002.  import tty
003.  import termios
004.  import threading
005.  import time
006.  from rpi_rf import RFDevice
007.
008.  # Elegant shutdown
009.  def exithandler():
010.      termios.tcsetattr(sys.stdin, termios.
      TCSADRAIN, old_settings)
011.      try:
012.          rx.cleanup()
013.          tx.cleanup()
014.      except:
015.          pass
016.      sys.exit(0)
017.
018.  # Activate our transmitter and received
019.  tx = RFDevice(17)
020.  tx.enable_tx()
021.  rx = RFDevice(27)
022.  rx.enable_rx()
023.
024.  # Receiving loop
025.  def rec(rx):
026.
027.      print("Receiving")
028.
029.      lastTime = None
030.      while True:
031.          currentTime = rx.rx_code_timestamp
032.          if (
033.              currentTime != lastTime and
034.              (lastTime is None or currentTime -
      lastTime > 350000)
035.          ):
036.              lastTime = rx.rx_code_timestamp
```

often used to switch lights. These commonly use 433MHz and protocols that rpi-rf can understand.

```
cd ~/rfchat
python3 receive.py
```

Press buttons on the remote control. You're likely to see a list of numbers, repeating for error correction, that change with each button. Make a note of these and then send them out as follows:

```
python3 send.py [number]
```

You should hear a reassuring 'click' from the relay of the socket. Try switching it on or off.

▲ A close look at the transceiver. The larger board is the receiver and the smaller square board transmits

10 Make it your own

These 433MHz units add a range of possibilities to your Raspberry Pi projects at a very low cost. Not just home automation projects with controllable sockets, but also providing radio communication where WiFi isn't practical, such as high-altitude ballooning or unusually positioned sensors like flood monitors. IoT devices can use radio to deliver and receive any information. Now you can control sockets from your Raspberry Pi, you can link these up to any kind of event you can imagine. How about detecting your car coming home using a Raspberry Pi Camera Module and number-plate recognition, then switching on the house lights? ☑

DOWNLOAD THE FULL CODE:

⬇ **magpi.cc/mcxmKh**

```
037.        try:
038.            if (rx.rx_code == 13):
039.                # Enter/Return Pressed
040.                sys.stdout.write('\r\n')
041.            else:
042.                sys.stdout.write(chr
     (rx.rx_code))
043.            sys.stdout.flush()
044.        except:
045.            pass
047.
048.        time.sleep(0.01)
049.
050. # Start receiving thread
051. t = threading.Thread(target=rec,
     args=(rx,), daemon=True)
052. t.start()
053.
054. print("Ready to transmit")
055.
056. # Remember how the shell was set up so we can reset
     on exit
057. old_settings = termios.tcgetattr(sys.stdin)
058. tty.setraw(sys.stdin)
059.
060. while True:
061.
062.     # Wait for a keypress
063.     char = sys.stdin.read(1)
064.
065.     # If CTRL-C, shutdown
066.     if ord(char) == 3:
067.         exithandler()
068.     else:
069.         # Transmit character
070.         tx.tx_code(ord(char))
071.
072.     time.sleep(0.01)
```

Advanced Projects

144

145

147

148

Advanced Projects

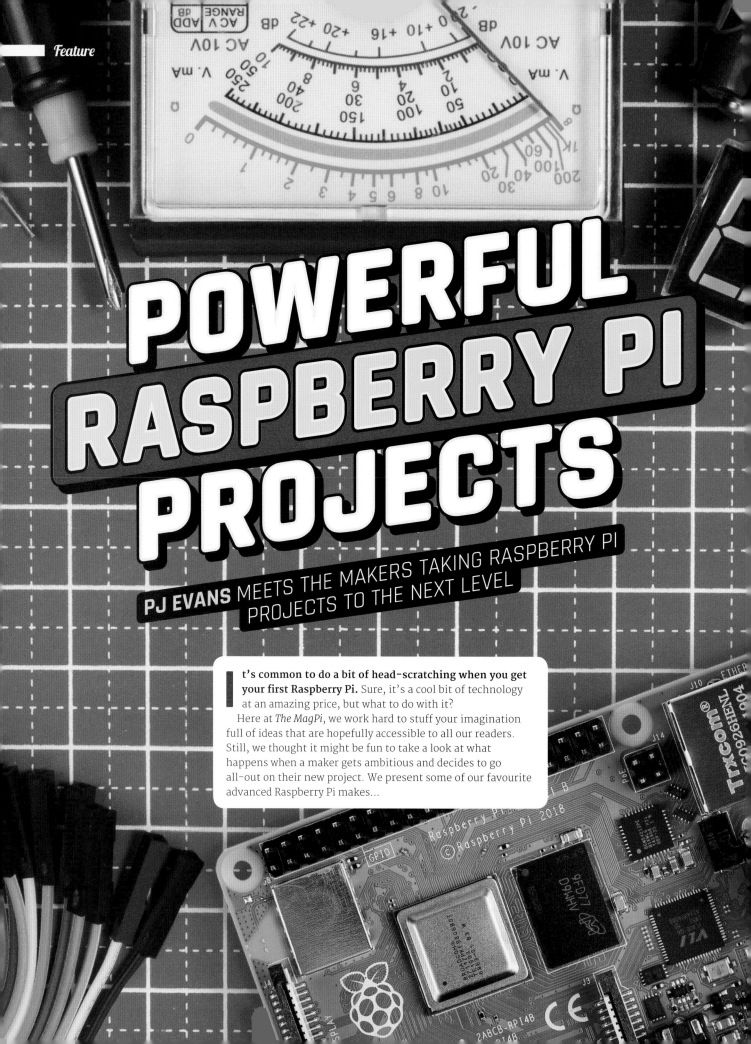

POWERFUL RASPBERRY PI PROJECTS

PJ EVANS MEETS THE MAKERS TAKING RASPBERRY PI PROJECTS TO THE NEXT LEVEL

I t's common to do a bit of head-scratching when you get your first Raspberry Pi. Sure, it's a cool bit of technology at an amazing price, but what to do with it?

Here at *The MagPi*, we work hard to stuff your imagination full of ideas that are hopefully accessible to all our readers. Still, we thought it might be fun to take a look at what happens when a maker gets ambitious and decides to go all-out on their new project. We present some of our favourite advanced Raspberry Pi makes...

PHOTO PROJECTS

Take amazing shots with Raspberry Pi

DROP PI

Maker **David Hunt** | **magpi.cc/fvNmeA**

Take incredible shots of water droplets, using a Raspberry Pi as a controller for a solenoid valve and camera trigger. The valve is hooked up to the GPIO pins and a small piece of code opens the valve and triggers the camera. The code is timed for a valve 40 cm above the surface of the water. It's a great example of how Raspberry Pi can be used to control an environment and camera, plus a good excuse to learn how to control valves.

STEREOPI

Maker **Eugene Pomazov** | **magpi.cc/pGvBAp**

Since 2014, Raspbian has offered built-in support for stereoscopic photography. With two cameras attached to a Raspberry Pi, you can create 3D photographs and record 3D video. You'll need a Raspberry Pi Compute Module (which has support for two Camera Modules). The compact and light nature of StereoPi makes it particularly useful for attaching to drones and robots.

ADVANCED GAMING BUILDS

Strictly for fun, these masterful makes put a smile on your face

HEVERLEE SJOELEN

Maker **Grant Gibson** | **magpi.cc/DQiGwQ**

When Belgium beer brand Heverlee approached prolific maker Grant Gibson for promotional ideas, he was reminded of Sjoelen, a shuffleboard game popular in Germany and Belgium. The result was a Raspberry Pi-powered physical game and vending machine mash-up that dispensed cold cans of beer to the winners.

THE CLAW

Maker **Ryan Walmsley** | **magpi.cc/aqtCxF**

Ryan's ever-popular claw machine is often seen at Raspberry Pi events throughout the UK. An upcycled bar-top 'grabber' game, this one can be played over the internet. Use your computer or mobile phone to try to grab Babbage the Bear (gently) as the results are live-streamed to you.

OUTRUN BAR-TOP

Maker **Matt Brailsford** (aka Circuitbeard) | **magpi.cc/TjjVMH**

What separates Matt from the crowd is his exquisite attention to detail. This OutRun Deluxe bar-top features fully working controls, such as gear shifting and a steering wheel. Add the pedals, repurposed from an wheel controller, and custom bodywork and this is a classy project.

CREATIVE PURSUITS

Physical computing and the arts are a very good match, and here's why

ETCH-A-SKETCH ART WITH PYTHON

Maker **Sunny Balasubramanian** | **magpi.cc/UvGhfe**

A classic toy featuring in many people's childhoods, the Etch-a-Sketch allows you to draw pictures without making a mess, then wipe the board with a satisfying shake. Sunny used servos and a Raspberry Pi to take control of the dials, then used edge-finding image filters to create an Etch-a-Sketch 'camera'.

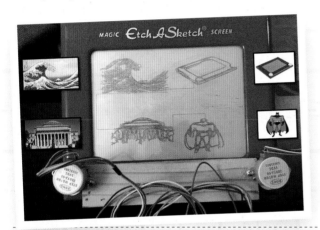

CUBERT

Maker **Lorraine Underwood** | **magpi.cc/dmSCSG**

A beautiful 8×8×8 structure of ping-pong balls, each containing a NeoPixel to create amazing colour patterns, and you can even play 3D Pac-Man. Of particular interest is Lorraine's blog and talks on the project, where she is brutally honest about the difficulties of making such a unique object.

NETWORK KNITTING MACHINE

Maker **Sarah Spencer** | **magpi.cc/iSVNEm**

One of the must-see exhibits at 2018's Electromagnetic Field camp was 'Stargazing' by Sarah Spencer. This map of the universe measures a colossal 4.6 by 2.8 metres and was knitted on a hacked Brother mechanical knitting machine, controlled by an Arduino and Raspberry Pi. It's even networked.

TRUE COMPACT CAMERA

Maker **Martin Parker** | **magpi.cc/dpbFJH**

Raspberry Pi Zero can fit into some tiny places, but how about a vintage 110 format compact camera? Martin replaced the internals with a Raspberry Pi Zero and Camera Module. Don't worry about film running out when your camera can take thousands of high-res snaps.

PI CLOCK 2

Maker **Tim Richardson** | **magpi.cc/dbqYPu**

Tim's clock is made up of two 64×32 pixel displays and displays information being relayed from a Raspberry Pi-controlled weather station. It features some clever energy-saving extras, such as a motion sensor to only update the screen when someone is in the room. It's been to Parliament as part of the one-millionth Raspberry Pi celebrations.

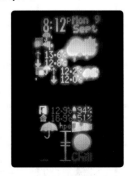

TAKE TO THE SEAS IN
YOUR AUTONOMOUS YACHT

MAKER

Al Coventry
(Coventry
University)

Left to Right: Balazs
Bordas, Mark Tyers,
Sergiu Harjau,
Shahzad Haider.

magpi.cc/uxjfhV

This is a wind-powered craft, so the sail is moved by motor

'The Rabbit', complete with Raspberry Pi Zero

The project's Raspberry Pi Zero is connected to a host of sensors to determine direction and speed

AI Coventry is making serious progress with autonomous vehicle technology. Sergiu Harjau and team entered their aquatic vehicle, 'The Rabbit', in an autonomous boat challenge in China. We asked him all about it.

What inspired you to build a self-sailing boat?

I first started having an interest in autonomous vehicles when I had to choose a project for a second-year module. I first built an autonomous RC Car, driven by Raspberry Pi Zero[…]. That got some traction in the university and then a lecturer offered me a spot on the autonomous boat team in Finland. We use the project as a way to broaden our skill set, both from a software standpoint but also when it comes to electrical engineering, and so far it's been working wonderfully.

What challenges did you face?

Autonomous vehicles are a bit like chess in some ways. It's very easy to understand how it's all meant to work, but it's really hard to go 'deep' and create beautiful systems which work without a single flaw[…]. In China, our biggest challenge which we didn't foresee was the weather. The humidity and extreme heat rendered some of our sensors faulty, spitting out random data at unpredictable times. Even still, we pursued our goals and in the end managed to fix some of the issues and came home with a pretty good result.

Are you happy with the outcome?

Yes, in our latest trip we did way better than our past ones, but even still we weren't perfect. We're very happy to call it a learning experience and go from there. On the flip side, we were very organised, more prepared than any team out there if I'm honest, and that allowed us to quickly fix our issues when we needed to. In the end, we managed to get third prize, and we're very happy with the result.

Any improvements planned?

We're going to be looking at spending a little bit extra on our compass sensor to ensure it doesn't get de-calibrated as often as it did in China. We suspected there were power lines under the lake, and that didn't help our autonomous sailing.

What plans do you have for your next vehicle?

Since autonomous vehicles and embedded systems are two of my favourite pastime activities, my next big project will again bring the two together. I'll be helping my lecturer Dr David Croft to deliver a hardware–software platform for a new master's course next year: 'Connected autonomous vehicles systems'. We're planning on building an RC car with capabilities to become autonomous on an ROS software interface. It's not going to be easy, to say the least, but I hope that through my other projects I have managed to gain the necessary skills to pursue yet another interesting endeavour.

A touch enclosure was required to withstand the rigours of space

All parts are 'commercial off-the-shelf', including the official Raspberry Pi Camera Module

The project's Raspberry Pi Zero required no modifications

TAKE PHOTOS
FROM SPACE!

MAKER

Surrey Satellites & University of Surrey

The Surrey Satellites team (below) with DoT-1, their 'Demonstration of Technology' satellite designed to show off new avionics complete with Raspberry Pi on board.

sstl.co.uk

Space is hard. Space, when you're not SpaceX or NASA, is extremely hard, but that didn't stop Surrey Satellites. Having secured a slot on Soyuz, they launched the DoT-1 satellite, which had a Raspberry Pi and camera on board supplied by the University of Surrey. It had a simple task: take a photo from orbit using commercially available off-the-shelf parts. We spoke to Surrey Satellites' Director of Engineering, Rob Goddard.

What inspired this project?

Whilst the primary objective of the DoT-1 (Demonstration of Technology) mission was to fly the company's next-generation avionics, there was space for some additional experimental payloads, hopefully stimulating the interest of our younger engineers. One of those experiments, designed and implemented in conjunction with the University of Surrey Space Centre, was to

> **❝ There could be some credible applications for low-cost computers and cameras of this type ❞**

capture an image from space using a commercial-grade Raspberry Pi Zero computer and camera, store the data, and downlink it via a new data handling system on board the satellite.

What challenges did you face?

It was a surprisingly easy project! We performed some screening tests on three Raspberry Pi Zero computers to select the best performing over-temperature, and then packaged the computer and camera into a metal box. The standard camera lens was changed for a fish-eye lens. The remaining electronics were completely untouched.

Are you happy with the result?

We were certainly pleasantly surprised by the quality of both the still imagery and video capture from Raspberry PI Zero and camera. There could be some credible applications for low-cost computers and cameras of this type. We're considering flying them as inspection cameras to confirm deployment of solar panels, or to view robotic arm movement.

RADICAL ROBOTS

We couldn't leave our little robotic friends out, now could we?

ROBOT DINOSAURS

Maker **Dr Lucy Rogers** | **magpi.cc/QPHdBc**

It's not often dinosaurs are disappointing, but when the animatronic dinosaurs of the Isle of Wight theme park Blackgang Chine kept failing, Lucy came to the rescue. Retrofitting the bespoke mechanics with off-the-shelf parts controlled by a Raspberry Pi, she made them easier and cheaper to maintain, with more movement options.

SOCCERBOTS

Maker **Neil Lambeth** | **redrobotics.co.uk**

Neil wanted to show kids that robotics can be more than just a single 'bot moving around, so he developed the SoccerBots, a pair of remote-controlled, ball-firing robots that try to score goals against each other. They've inspired hundreds of kids around the UK.

BIOHEX

Maker **Harry Brenton** | **magpi.cc/eynBqs**

Hydroponics is all the rage. Growing plants without soil produces amazing results, but requires careful monitoring and care. Harry's BioHex is a 3D-printed modular plant-growing machine that uses a Raspberry Pi to analyse the environment, operate the air pump, and provide lighting control.

OPEN-SOURCE MARS ROVER

Maker **NASA Jet Propulsion Laboratory** | **magpi.cc/yhXYKp**

Want your own Mars Rover? Of course, you do. After a successful educational outreach programme demonstrating a small version of its real rovers, NASA's JPL created a new robot, ROVE-E, that is made of off-the-shelf parts and runs open-source software on a Raspberry Pi. It costs around $2500 to build.

INCREDIBLE AI PROJECTS

Build something super-smart with Raspberry Pi

GOOGLE CORAL

Maker **Google** | **magpi.cc/coral**

Coral is a range of AI products and projects made by Google. The latest star is the USB Accelerator. This dongle adds a Google TPU to Raspberry Pi, which rapidly boosts real-time classification – all local on the Raspberry Pi. Check out Teachable Machine in *The MagPi* 79 (**magpi.cc/79**).

TENSORFLOW CUCUMBER SORTER

Maker **Makoto Koike** | **magpi.cc/fDeLLK**

It's lovely to see one maker put AI tech into practice to solve a problem. Makato's father grows cucumbers and straight ones with lots of prickles command a high price. Makato trained an AI robot to spot and sort them. Perhaps there are a number of production lines that could benefit from a little AI.

THERE'S WALDO

Maker **Matt Reed** | **magpi.cc/LaGxbn**

Where's Wally? (Or Waldo as he's known in the US.) Matt Reed's There's Waldo robot is an irreverent demonstration of machine-learning. There's Waldo uses OpenCV to extract all the faces from the page, and then sends them to Google Auto ML Vision service. This locates the striped one and sends it back to a Raspberry Pi, which points him out using a uArm metal arm.

WHOA! WHAT IS THAT?

Some projects are just so 'out there', they defy categorisation

HIGH-ALTITUDE BALLOONING

Maker **Dave Akerman** | **daveakerman.com**

The world's best computer is no stranger to space, often hitching a lift on the ISS. If you'd like to send a Raspberry Pi skyward, it's easier than you think. Dave sends Raspberry Pi Zero computers up to 100,000 ft (30,480 m) in the air using helium balloons, taking amazing photos. His comprehensive blog shares an immense amount of knowledge.

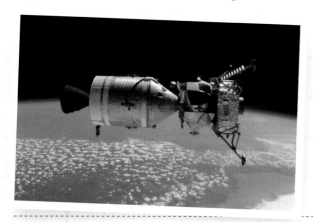

3D SCANNER

Maker **Richard Garsthagen** | **pi3dscan.com**

This project gets a special place on our 'Whoa!' list simply for 98 Raspberry Pi computers in use. Split over 19 poles, they provide a high-resolution 3D model of anything placed within the scanning area. Richard has not only written a great series of posts as the project has evolved, but has open-sourced many of the plans and code used.

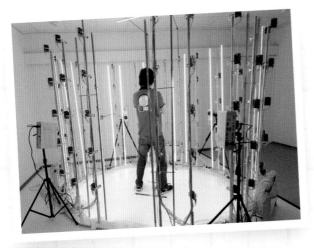

Warning!
High Voltage

This project uses potentially dangerous levels of electricity

MUSICAL TESLA COIL

Maker **Derek Woodroffe** | **magpi.cc/WhJQRq**

A Tesla coil as a musical instrument? Why not? Derek is famous in the community for his high-voltage antics. Here, a Raspberry Pi Zero can load MIDI files and convert each note to information that is sent to two coil driver boards. The result is your favourite tune rendered at 200 volts.

FLAPPY BRAIN

Maker **Albert Hickey** | **magpi.cc/jhspVm**

Mattel's 'Mind Flex' toy appears to read your mind by measuring your brain wave activity. Egham Jam organiser Albert Hickey added an Arduino and Raspberry Pi to decode the output so you can play Flappy Bird. Move the bird down by thinking hard, then clear your mind to go up!

FLOOD NETWORK SENSOR

Maker **Ben Ward** | **flood.network**

Citizen-scientist Ben Ward has invented a cheap, serviceable alternative to expensive flood-monitoring systems for his home town of Oxford, using Raspberry Pi computers. The system bounces sound waves off the water surface to calculate the level. The reading is then relayed by radio (LoRa) to a central database. The project is now spreading across the UK.

SPIN UP A DIGITAL ANIMATION

An Arduino helps track the rotation of the zoetrope

Each screen shows a single frame that is part of the animation

Brian designed a custom driver board for each screen so a single Raspberry Pi can refresh every one

MAKER

Brian Corteil

Brian is an award-winning robot maker, active member of the Raspberry Pi community, and self-titled 'head meat-bag' of Coretec Robotics.

magpi.cc/LjJFFO

Brian is a regular at Raspberry Pi events up and down the UK and is most often seen with his Pi Noon balloon-battling robots or FacePlant, the two-wheel balancing creation he entered in this year's Pi Wars. There is another one of his builds that caught our attention, a digital zoetrope (an early animation machine) that threw up some real technical challenges, in particular trying to drive twelve screens from one Raspberry Pi. The images on the screens can be updated in real-time when the zoetrope is spun.

What inspired you to build a zoetrope?
While I was researching the images of Eadweard Muybridge and the history of moving images, I was reminded of the zoetrope. I had a crazy idea that I could make a digital version bringing Eadweard Muybridge's images to life.

What challenges did you face?
I had to control twelve screens on a single Raspberry Pi and design a circuit to be able to select each screen. Then I needed to write software to render images for the screens, modifying the driver software to upload an image in four blocks.

Are you happy with the result?
I'm pleased with the way the Digital Zoetrope turn out after I changed the shiny black acrylic to matt black and the wiring to black. It ended up being as I first imagined it would be.

Any improvements planned?
To make it more interactive and be able to import cells by taking a photo of a hand-drawn sheet of a one-second short film.

Are there any more zoetropes in your future?
Well, I have two projects planned. Hopefully, I will be making a second larger version of the Digital Zoetrope, using e-paper displays for Electromagnetic Field 2020 camp as an art installation, using twelve Pi Zeros and a Raspberry Pi 4 networked together. And of course, my annual robot build for Pi Wars. **M**

Build a low-cost
wheeled robot

You want to build a robot without breaking the bank. Let's see what parts we need, where they fit, and how to keep the cost down

MAKER

Danny Staple

Danny makes robots with his kids as Orionrobots on YouTube, and is the author of *Learn Robotics Programming*.

orionrobots.co.uk

To make a robot, be it a wheeled rover, flying drone, factory robot, or autonomous spacecraft, you will need common classes of components. We'll discover what they are for, focusing on those needed for a wheeled robot.

We'll look at what options there are for the components, and how we might be able to save money. We'll go through the trade-offs needed for these options, the tools you might need, and their relative difficulty.

Any robot starts with a computer to run code, using sensors to collect data about the world. There are output systems to drive motors and actuators to affect the world. It needs power systems to get the right voltage and current to the right parts. The robot will need mechanical parts for the motors to drive, along with connecting the sensors and a body holding it all together.

▶ These parts are ready to be built into a lunchbox chassis with plastic gear motors and plastic wheels. We will go into more detail on turning a lunchbox into a robot

01 An overview of robot parts

To make a wheeled robot, you are going to need some common part types to make it work:

- A chassis or body to hold everything together. You will need brackets for sensors eventually, too.
- Wheels and motors to drive them. This includes balance wheels or castors.
- A main controller to run your code: Raspberry Pi.
- A motor controller or driver to connect your Raspberry Pi safely to outputs.
- Batteries and power regulation for your electronics.
- Sensors to get data from the real world, like distance sensors and a camera.

02 Going low-cost

To go low-cost, you going to have to be a little creative. This will mean substituting parts, or finding parts that may not be the obvious choice. You will be able to save by shopping around, and waiting for parts that will take longer to ship will usually reduce cost.

Having parts pre-soldered or ready-made usually adds quite a lot to their cost, so be prepared to solder things together for the electronics, and to bodge or repurpose things for the body.

Robots can be made of anything, although we're not advocating it; we've even seen a robot made with vegetables for a body and wheels.

03 Raspberry Pi

The robot needs a Raspberry Pi to run your code. How would you save here? Well, the first cheapest Raspberry Pi is the one you already have!

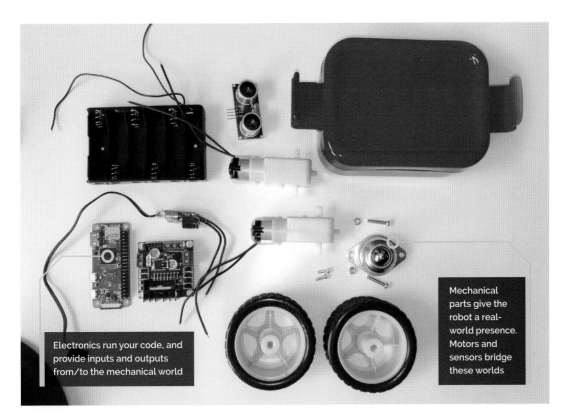

Electronics run your code, and provide inputs and outputs from/to the mechanical world

Mechanical parts give the robot a real-world presence. Motors and sensors bridge these worlds

Top Tip

Get creative and repurpose

Be on the lookout for unused brackets and plastic shapes that might be handy to use in robot builds.

Make sure it's one with wireless LAN and a 40-pin header if you can (Raspberry Pi 3, 4, and Zero W models are all good choices).

Our favourite low-cost, and low-space, option is Raspberry Pi Zero WH (**magpi.cc/zerowh**). This is the smallest model, with wireless LAN and a GPIO header pre-soldered to the board. Since a robot isn't often connected to the screen, the lack of DSI port shouldn't be a problem. If you have a Raspberry Pi Zero W, then soldering pins to the GPIO header is a fun project (**magpi.cc/soldering**).

04 The chassis or body

The chassis holds the robot together. It's a fundamental decision on how your robot is made. Options for a wheeled chassis are:

- **Easy**: A laser-cut chassis – these are cheap to buy and easy to work with. They have space to extend the robot. They can be flimsy and do break, but are the simplest option and do not require much in the way of tools and time. If you go for a kit, a two-motor variant is advisable. Not as much fun as the lunchbox option.
- **Easy**: Adapt a lunchbox. This does require a little measuring and drilling of holes, but is still quite an easy option in terms of construction. You will need to choose smaller parts to fit in the lunchbox.

" A good first robot is the lunchbox robot. It's a good compromise of saving cost and complexity "

- **Intermediate**: Cut a chassis from wood or sheet material – this requires access to woodworking tools and CAD/drawing skills.
- **Intermediate**: The toy hack – one of our personal favourite options is taking a cheap motorised toy and swapping its electronics for a Raspberry Pi and motor board.
- **Hard**: The least cost for a chassis is not to have one and go for a free-form robot, strapping motors directly to control boards and batteries. This is a lot of fun, but takes experience and practice. Cable ties may be all you need to buy for this.

A good first robot is option two, the lunchbox robot. It's a good compromise of saving cost and complexity, it's fun, and has a little bit of character. So that's what we're making here.

05 Not reinventing the wheel

Drive wheels will be attached to your motors. For a low-cost robot, two driven wheels is a good number.

Unless trying to make experimental robots, plastic wheels with rubber tyres make the best drive wheels and are cheap and readily available.

- Both wheels need to be the same diameter.
- They need good grip.
- Axles should be aligned in the centre, and not slip.

Making your own is possible, but inadvisable. Instead, we recommend buying plastic wheels with tyres (search for this phrase with 'robot'), available for less than £5, and often with the motors for not much more.

06 Castor wheel or ball

A castor wheel, ball, or skid is mostly on the robot to balance it, without causing too much friction. You can get away with a simple bottle lid facing down here or, for a better cheap hack, half a ping-pong-ball. You can buy robot castor wheel or roller ball assemblies for under £2.

The important thing is that this component does not introduce friction, and can easily be attached firmly to the robot.

07 Types of drive motors

What kinds of motors are there?

- DC motors simply spin. They are cheap, but are easily stopped with any load.
- Stepper motors move in 'steps', a fraction of a full turn. They are not cheap, but can be salvaged from old printers and scanners.
- Servo motors can be moved to a particular position but not make a full revolution. They can be controlled directly from a Raspberry Pi. These can be modified for continuous rotation, but this can be complicated or expensive.

- DC gear motors combine a gearbox with a DC motor to drive heavier loads. They are cheap and easy to find. We recommend this option.

08 Choosing the gear motors

Buy motors with a gear ratio above 40:1. Although they'll never match perfectly, get them in pairs, as motors that look similar may not have the same speed.

They come in plastic and metal geared flavours. Metal tends to last longer but cost more. Plastic gear motors are cheaper, but larger and not as sturdy. In most of our robots, plastic motors are fine.

90-degree motors fit better in a limited space. Ensure the axles match your wheels. Adapting axles adds complexity and cost.

Gear motors can be salvaged from an electronic toy like an RC excavator or tank.

09 The motor controller

This connects your Raspberry Pi to the motors. Ensure it can control DC motors.

- Buy a Raspberry Pi HAT designed for motors. This is more expensive than other options, but may have additional functionality like logic shifting for sensors, servo motor control, or power regulation. It'll cost £15–£25.
- A DC motor control breakout module. These should have two channels. The cheaper options tend to be based around the L298N or DRV8833 chips, good enough for our purposes. Should be under £8. A simple and reliable option.
- Someone with more electronics knowledge could construct their own H-bridges, but this probably won't cost less than the DC motor modules.

10 Power up

Robots need power for the motors and electronics. A Raspberry Pi needs a smooth 5 V to run, with upto 3 A capacity. Motors introduce noise that could interfere with your Raspberry Pi.

- Separate power, using a USB power bank for Raspberry Pi and other batteries for motors. Power banks can be pricey, but we've seen smaller ones given away at shows.
- A single set of batteries with a regulator like

▼ These are some motor examples. From left to right: a DC motor, a plastic gear motor, a metal gear motor, and a servo motor

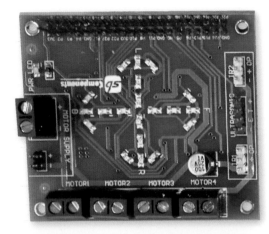

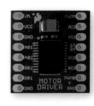

A selection of motor controllers. The DRV883 and L298 are small and cheap, with the larger types having more features

a UBEC (universal battery eliminator circuit). These can be bought to provide 5 V and more than 2 A. A UBEC can be found for around £6.
- LiPo/Li-ion are expensive and tricky, only recommended for experienced builders.

11 Making sense

For more interesting code, a robot needs sensors to detect things. Those that don't require extra conversion will save space and a little money. Be prepared to solder on headers.

An HC-SR04P is a cheap way to measure distance. Line-tracking sensors let the robot follow lines.

A camera can be used to follow lines and detect certain objects optically.

An optical flow sensor looks a bit like a camera and lets you measure x-y movement, but cannot tell when you have turned.

Motion sensors can tell you acceleration, rotation, and change in heading. 9DOF types are now around £12.

12 Fasteners and wiring

For fasteners, the 'standoff kits' found online are great; get a selection of M2, M2.5, and M3 including the bolts, nuts, and spacers.

Cable ties are cheap and handy. They can tidy up cabling but can also strap robot parts together.

> ❝ For more interesting code, a robot needs sensors to detect things ❞

Hot glue is handy for stopping things rattling around; however, don't use it for load-bearing connections. Double-sided sticky tape, foam, and sticky tack can also be used this way.

Jumper wires are used to get signals and power between your components. We recommend getting male-to-female, male-to-male, and female-to-female jumper cable selections (for breadboarding). These will connect Raspberry Pi to most sensors and modules.

13 Tools

A hot glue gun has already been mentioned. Some kind of hand drill and a way to clamp parts will be needed for any custom building. DO NOT hold a part you are drilling in another hand!

Also useful is small screwdriver set, and miniature spanners for fastening parts. We recommend needlenose pliers, side cutters, a multimeter, and a soldering iron.

The best way to keep costs down is to borrow tools – if you have a local hackspace, makerspace, or can talk to a college about using their facilities, you won't need to buy all these tools. 🔲

Build a low-cost
wheeled robot

Use a lunchbox to build a cheap wheeled robot! Equipped with motors and Raspberry Pi, it is an excellent platform for robotic experiments

MAKER

Danny Staple

Danny makes robots with his kids as Orionrobots on YouTube, and is the author of *Learn Robotics Programming*.

orionrobots.co.uk

The first part of this series showed parts used to make low-cost wheeled robots. You can get a lunchbox for very little money, and they make for sturdy small robots.

Their size leads to some constraints. Raspberry Pi Zero W (or WH) fits well in these spaces. The L298 controller is a reasonable balance of cost, size, and ease of use.

A robot builder needs access to a few tools like a ruler, paper, pencil, a drill, some screwdrivers, and a vice/clamp. This build uses consumables like jumper wires, screws, AA batteries, and a standoff kit with M2, M2.5, and M3 types. A hot-glue gun is also handy.

You'll Need

- Plastic lunchbox – minimum 90 mm wide × 120 mm long × 60 mm high

- 2 × Gear motors with wheels
magpi.cc/nhoAbj

- Battery box – 6 × AA
magpi.cc/DPFaEJ

- 5 V 3 A UBEC
magpi.cc/hZohEg

- Ball castor
magpi.cc/bfALSE

- L298N module
magpi.cc/EDKvrL

- Nylon screw and standoff kit
magpi.cc/XaRfgp

01 Getting familiar with components

Examine the motors. Sticking out of the sides are the axles. On one side, close to the axle, is a small knob which can help lock the motor into position. The top and bottom of the motor is flat. The

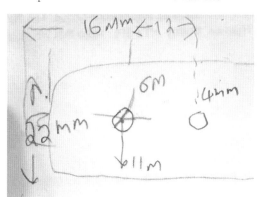

▲ Engineering sketches are essential in robotics. The sketch can be rough with dimensions and notes. Sketch lots! This is the side with motors

motors should have connection wires on them, on the opposite side to the knob.

The motor controller has a large metal heat-sink sticking up above the board.

Raspberry Pi GPIO pins should be present and go in face up so they can be used.

A lunchbox is made of thin rigid plastic. It probably has rounded corners. Take measurements using the flat edges.

02 Planning and test fitting

Test-fit the parts to check the lunchbox is big enough. Make a simple plan for where parts could go.

Put the motors in the box, as shown in **Figure 1**, with their axles to the rear. Fit your Raspberry Pi Zero in the front space and the motor controller between the motors. The UBEC (which is a DC voltage regulator) sits loosely in the box.

The battery box goes in the top – inside if it can; outside if it interferes with the heat-sink.

The wheels would go on the motors and the castor at the front under Raspberry Pi. Take a photo and remove it all.

03 Sketches for the motors and castor

Sketches enable you to place holes and components before making any cuts. See orthographic drawing tips at **magpi.cc/tZKdQV**.

Find or start a sketch of the motors. Write in dimensions for their size, axle diameter, and location, and the positioning knob on the side.

Sketch a side of the lunchbox, with holes to put motors flush with the inside bottom of the

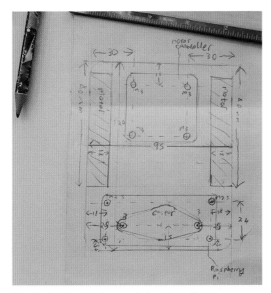

▲ This sketch is of the bottom of a lunchbox with holes and dimensions. Yours will be slightly different. It's on the back of an envelope

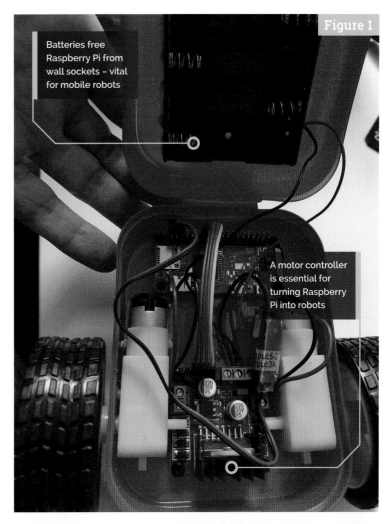

Batteries free Raspberry Pi from wall sockets – vital for mobile robots

A motor controller is essential for turning Raspberry Pi into robots

Figure 1

lunchbox. Include dimensions to show relative positions of holes for the axle and knob.

Sketch the bottom of the lunchbox and measure the two mounting holes for the castor to go in the front middle.

You should have two lunchbox sketches with carefully checked dimensions.

> ❝ Sketches enable you to place holes and components before making any cuts ❞

04 Measuring for the electronics

The test-fit photo (**Figure 2**) is a rough guide on positioning things in the box, but sketches with dimensions have detail on where components should go.

Raspberry Pi has drawings with dimensions at **magpi.cc/HdffiN**. Add mounting holes for it to your lunchbox bottom sketch near the front, with dimensions for diameter and position.

Measure then sketch the motor controller's position and holes for the controller on the lunchbox bottom sketch.

Carefully check the bottom sketch. It should have holes for the castor and both electronic boards.

Sketch the top of the lunchbox, where the AA battery box goes.

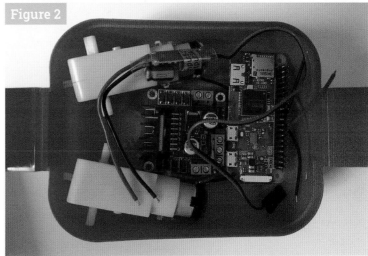

Figure 2

▲ **Figure 2** A fundamental step in robot creation is test-fitting where items might go! Use the real components or drawings of them

05 Drill the holes

Double-check sketch measurements. Use a fine marker pen to measure and make crosses as hole guides on the lunchbox.

Clamp the box firmly with a vice, ensuring the surface to drill is facing upwards. *Do not* try to hold the box with hands when drilling it!

When making holes, the part may flex, crack, or spin. The drill bit can 'wander' from its target. To reduce this, start with the smallest bit to make a pilot hole. Follow through with progressively wider drill bits to the desired diameters. You'll need to reposition the lunchbox to drill different areas.

❝ Don't tighten nuts past finger-tight, as this can damage the bolts or the parts held by them ❞

06 Fit the motors and castor

Line up the motor axles and knobs with their holes, then push the axles through. If any holes feel tight, drill them out a little more. Make a hot-glue line between each motor and the lunchbox to hold it, being careful to avoid getting glue on the axle.

Bolt the castor in place using M3 nylon bolts and nuts with the threads facing outward. Start with two opposite corners if it has four bolts. These bolts go under Raspberry Pi; to avoid short circuits, they must not be metal bolts.

07 Wiring the motors

The motor board may be harder to access when bolted down, so it makes sense to make the motor connections at the sides first.

Start by loosening, but *not* wholly unscrewing the connection terminals. Look at the circuit. Push the wires for each motor into the two terminals for that side, and then gently but firmly screw down the terminal so it holds the wire.

▼ The website **pinout.xyz** is an excellent reference for Raspberry Pi GPIO pins. Use this to help when wiring the robot. Black pins are ground

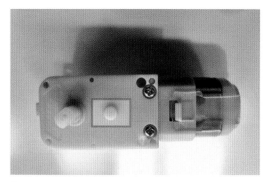

▲ Next to the axle, a motor has a knob (highlighted) to help align it. Include these in sketches as these will need holes too

It's a good idea to tug gently on the wires to check they are screwed into the terminal securely.

08 Fit Raspberry Pi and motor controller

Line up the motor controller with its holes. It may need to slot under the motor axles. Push M2.5 bolts through two opposite corners to line it up, and loosely put nuts on these. Put in the other two corners and tighten them all up.

Don't tighten nuts past finger-tight, as this can damage the bolts or the parts held by them.

For Raspberry Pi, put M2.5 standoffs thread first into the four holes and tighten nuts to them. Place your Raspberry Pi Zero W over the standoffs. Use the opposite corners method again. Leave these quite loose so your Raspberry Pi can be taken out.

09 Fitting the battery box

Using the sketch made of the top of the lunchbox, line up the battery box – on the inside if there is space; on the outside if it would stop the lunchbox closing. An extra hole is needed for cables if it's outside.

Screws could be used, but flat countersunk types are needed, making hot glue an easier option.

Make lines of hot-melt glue between the battery box and lunchbox around the sides of the battery box. There should be no glue inside the battery box, and it should be held firmly once the glue has set.

10 Wiring the batteries

The battery box connections and UBEC are wired in together. The UBEC has two sets of

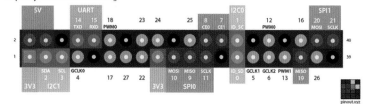

Raspberry Pi GPIO BCM numbering

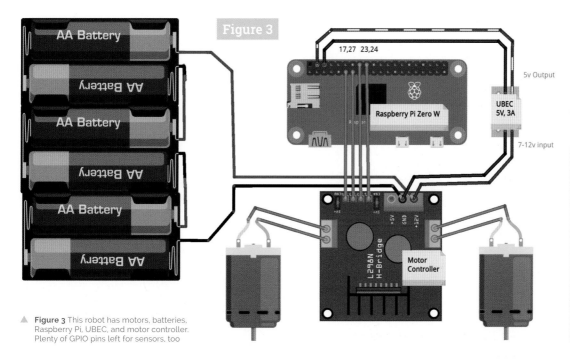

Figure 3

17,27 23,24

Raspberry Pi Zero W

5v Output

UBEC
5V, 3A

7-12v input

L298N
H-Bridge

Motor
Controller

▲ **Figure 3** This robot has motors, batteries,
Raspberry Pi, UBEC, and motor controller.
Plenty of GPIO pins left for sensors, too

Top Tip

Part sketches on the internet

Searching the web for part dimension pictures provides useful sketching start points. Print them and draw more on them, or use them for reference when making other sketches.

connections. The input end has thicker wires with stripped ends, and the output connector has a three-pin push-fit connector.

Wire the battery positive (red) and UBEC input positive (red) together into the motor controller +12 V or Vin terminal. It helps to twist the two bare ends together before pushing into the terminal and screwing down.

Wire the battery negative (black) and UBEC input negative (black) together into the motor controller GND terminal.

If you want to add a switch, put it between the batteries and the UBEC / motor controller.

---

11 Wiring Raspberry Pi

Use the circuit diagram (**Figure 3**) with Raspberry Pi GPIO pinout as guidance for this step. Physical pin 1 is closest to the microSD slot on Raspberry Pi Zero W boards.

The UBEC three-pin push-fit connects to GPIO so that the empty slot and red wire go to the 5 V pins, with the black wire going to the ground pin.

Use female-to-female jumper wires to connect the Raspberry Pi to the motor controller pins. Connect Raspberry Pi GPIO 17 and 27 pins to motor controller IN4 and IN3 pins.

Connect Raspberry Pi GPIO 23 and 24 pins to motor controller IN2 and IN1 pins.

12 Finishing the robot

To finish assembly of the robot, push the wheels onto the axles.

Line up the wheels so the flat parts of the axles line up with the flat edges in the axle holes. They can require quite a bit of force, so push on the motor and the wheel to deliver it. Do not try to push the wheels on without supporting the motor.

You can now also place the lid on top of the robot. If the lunchbox isn't quite tall enough, the lid might rest on top – do not force it down.

This robot is built, wired, and waiting for code! 🄼

▼ The components for the robot, including a battery holder glued to the lunchbox lid

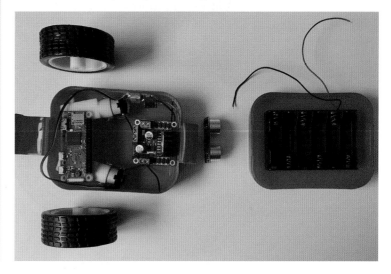

Build a low-cost wheeled robot

A wheeled robot isn't much fun when it's stationary. Using the motor controller and code, it's time to get the robot moving!

MAKER

Danny Staple

Danny makes robots with his kids as Orionrobots on YouTube, and is the author of *Learn Robotics Programming*.

orionrobots.co.uk

You'll Need

- Lunchbot
- Set of 6 × AA batteries (nice and fresh, alkaline)
- 16GB microSD card
- microSD card reader
- WiFi network

T he previous part of this series showed how to build a robot using a lunchbox as a chassis (call it Lunchbot!). While that robot looks smart, and like it might work, it's not going to do much without code to do something on it. In this part, we'll get your Raspberry Pi ready for robotics and write some simple code to move the robot.

You'll set up the microSD card for running code over wireless LAN, add tools to connect to the robot's motor board, and write code to instruct the motors to turn.

01 Prepare the microSD card

When the robot is moving, Raspberry Pi can't be tethered to a screen or keyboard. A 'headless' configuration is needed. Headless means designed to be operated via the network. A NOOBs or Raspbian desktop image is unsuitable for this, so use Raspbian Lite.

Download a Raspbian Lite image from **magpi.cc/raspbian** and burn it onto the microSD card. Etcher (**balena.io/etcher**) is recommended

for this. When it is ready, eject the microSD card and put it back into the computer. Then follow the instructions at **magpi.cc/RhviuV** to configure WiFi (in the **wpa_supplicant.conf** file) and enable SSH.

02 Connect to your Raspberry Pi

The microSD card and USB power input on the robot's Raspberry Pi may not be accessible. If needed, loosen the bolts to get to it. Put the microSD card into the robot's Raspberry Pi.

You can power this Raspberry Pi from a USB cable during setup. When the lights stop flashing, Lunchbot's Raspberry Pi has booted and should be ready to contact.

Raspberry Pi has a guide on finding it on the network at **magpi.cc/kNOAbE**. It's called **raspberrypi.local** for now, but you can change this by following **magpi.cc/HZGBFR**.

Since it is headless, SSH is the right way to interact with the robot. See **magpi.cc/2sLqBmM** and **magpi.cc/ssh** for instructions on getting in.

03 Installing robotics software

This robot is programmed in Python 3, using GPIO Zero. Use SSH to access Lunchbot's Raspberry Pi, then install GPIO Zero with:

```
sudo apt-get update && sudo apt-get upgrade
sudo apt-get install python3-pip python3-gpiozero
```

Type **python3**. You should see something like:

```
Python 3.7.3 (default, Apr  3 2019, 05:39:12)
[GCC 8.2.0] on linux
Type "help", "copyright", "credits" or "license"
```

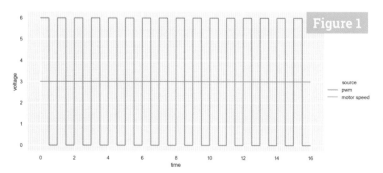

▲ **Figure 1** A PWM square wave in action, keeping a motor at a constant speed by turning the power on and off

Table 1

IN 1	IN 2	Motor
Low	Low	Off
Low	High	Spin Forward
High	Low	Spin Backward
High	High	Off

> for more information.
> `>>>`

Type `import gpiozero` to test GPIO Zero is ready. It should not show an error. Press **CTRL+D** to exit. Shut down your Raspberry Pi with `sudo poweroff`, remove the USB power, and bolt it in place.

04 Motor controller concepts

The motor controller has two halves. Each half controls a motor using two outputs, using power and three I/O input pins.

The motor outputs can handle higher currents and voltage than the inputs need. Connecting motors directly to Raspberry Pi would damage it.

The enable input pins turn on half of the controller. Labelled ENA for motor A or ENB for motor B, they are connected with a 'jumper wire' to a 5 V line (held high) to keep the half enabled. Each motor has two direction control pins: IN1 and IN2 for motor A; IN3 and IN4 for motor B.

05 Controlling speed and direction

The IN pins are used in pairs to control a motor. See **Table 1** for how the pins control the direction of the motor. You can substitute IN 3 and IN 4 for the behaviour on the other motor.

drive_forward.py

DOWNLOAD THE FULL CODE: ⬇ **magpi.cc/QkkmqW**

> Language: **Python 3**

```
001.  import gpiozero
002.  import time
003.
004.  robot = gpiozero.Robot(left=(27, 17), right=(24, 23))
005.
006.  try:
007.      # Robot actions here
008.      robot.forward()
009.      time.sleep(1)
010.  finally:
011.      robot.stop()
```

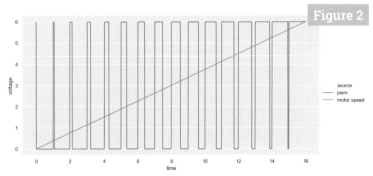

▲ **Figure 2** PWM can control the speed of a motor by changing the on/off ratio

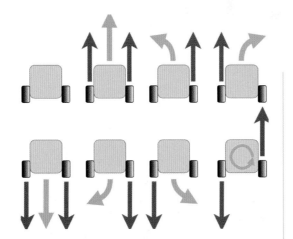

How a wheeled robot moves. The dark arrows show the direction of the wheels; the green arrows show the movement of the robot

This controls the direction, but what about speed? By turning the motor power on and off rapidly (in square waves), the ratio of time on vs time off changes the speed of the motor. **Figures 1** and **2** show how this works. This is known as pulse-width modulation or PWM. The GPIO Zero library can set the direction and the PWM speed.

Top Tip

Have plenty of driving space

The robot is going to move and needs space to do so. Don't test this on a tabletop!

06 Not driving off the table

Robots under test have a tendency to shoot off until they hit a wall or off a desk, and not in the direction you expect. Emergency stop measures are essential. First, ensure that motors are always turned off. Python's `try…finally` block applies in almost all cases. Even if the program crashes, it runs the `finally` code.

```
try:
    pass
    # Code that starts motors
finally:
    # Code to stop motors
```

The other measure is an emergency stop button. Python comes with such a button built-in: pressing the keyboard combination **CTRL+C** will stop the code running and enter the `finally` code.

07 Setting up robot code

The setup for robot experiments is in **drive_forward.py**. Use this as a template for robot code. Line 1 and 2 make the GPIO Zero and time libraries available by importing them.

GPIO Zero has a handy robot object to control two motors. At line 4, a robot object is created with pin numbers for the motor controller.

Lines 6 and 10 provide the stop mechanism from Step 6, with line 10 stopping the robot.

Between these lines, after the comment (starting with a '#') on line 7 is the code to make the robot do something. This code can be swapped/changed for different movement.

08 Driving forwards

In **drive_forward.py**, line 8 instructs the controller to drive both motors forward.

The delay on line 9 is needed so that the robot drives for a little time before stopping the motors. `time.sleep()` uses a time in seconds.

Put in batteries to switch to battery power. If you don't have a switch, use the last battery as one.

Use SCP to put this code on the robot or use Raspberry Pi to edit it. Put it in **/home/pi**. Give the robot space to drive and run the code:

```
python3 drive_forward.py
```

It should drive forward for one second and stop. Press **CTRL+C** if it does not stop.

09 Troubleshooting driving forwards

If running **drive_forward.py** shows errors, check everything in Step 3, and that the code is exactly as listed.

If one motor (or both) doesn't move, inspect the motor controller wiring. Ensure the enable jumpers are fitted. If the motors whine but don't move, try fresh batteries. If a motor goes backwards, then swap its pin numbers on line 4.

The robot may veer to one side; check the motors are glued in well, without friction between wheels, axles, and robot. Some veer is unavoidable without sensors. This is when loose screws in the robot will show up and need fixing.

10 Simple turns and driving backwards

The robot object in GPIO Zero makes it very easy to do this for the simplest movements.

Where the code has `robot.forward()` on line 8 of **drive_forward.py**, this can be replaced with `robot.left()`. It is only now you will find out if you have the motors the correct way around! If the robot drives to the right, then swap the **left** and **right** keywords on line 4.

The robot is likely to turn too far, so reduce the timer value in line 9, this can be decimals, so try

0.3 seconds here. On line 8, try `robot.right` and `robot.backward` too.

11 Chaining movements

So far, this code has been straightforward with a single movement. A robot can do a sequence of actions after each other.

The code listing **chaining_movements.py** shows how. By tuning the timings, this will drive in a shape such as a hexagon.

Line 8 makes a loop, so the action will be repeated. Lines 9, 10 drive forward for a little time. Lines 11,12 then turn left for a bit of time too.

Lines 13 and 14 then do a victory spin to the right. An excellent, fun way to end autonomous events if you can determine a victory condition.

12 Moving at slower speeds

A robot's motors can do more than just change direction: PWM can be used to vary the speed. The listing **speed_control.py** shows how.

Line 8 sets up a loop, in this case counting down from 10 to 4. The speed is set in line 9 using `robot.value`, with the left value setting the left motor speed, and the right value setting the right. Values go from zero to 1, so divide by 10. Negative values go backwards.

Lines 12 to 15 do some turns with more precise control than the `robot.left` or `robot.right` commands. These will be handy for sensors.

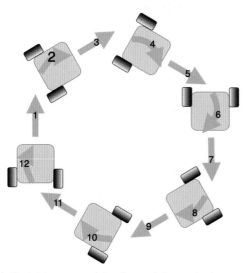

By chaining movements together, a robot can make shapes on the floor. Add a pen and it could draw them too!

chaining_movements.py

> Language: **Python 3**

```
001.  import gpiozero
002.  import time
003.
004.  robot = gpiozero.Robot(left=(27, 17), right=(24, 23))
005.
006.  try:
007.      # Robot actions here
008.      for n in range(6):
009.          robot.forward()
010.          time.sleep(0.5)
011.          robot.left()
012.          time.sleep(0.3)
013.      robot.right()
014.      time.sleep(1)
015.  finally:
016.      robot.stop()
```

> A robot's motors can do more than just change direction: PWM can be used to vary the speed >

speed_control.py

> Language: **Python 3**

```
001.  import gpiozero
002.  import time
003.
004.  robot = gpiozero.Robot(left=(27, 17), right=(24, 23))
005.
006.  try:
007.      # Robot actions here
008.      for speed in range(10, 4, -1):
009.          robot.value = (speed/10, speed/10)
010.          time.sleep(0.4)
011.      # Smaller turns with forward motion
012.      robot.value = (0.5, 1) # left
013.      time.sleep(1)
014.      robot.value = (1, 0.5) # right
015.      time.sleep(1)
016.  finally:
017.      robot.stop()
```

Build a low-cost
wheeled robot

Make a robot react to the world with sensors. Fascinating behaviours emerge with only a bit of code and electronics!

Danny Staple

MAKER

Danny makes robots with his kids as Orionrobots on YouTube, and is the author of *Learn Robotics Programming*.

orionrobots.co.uk

Over the previous three tutorials we've built a low-cost wheeled robot. Without any sensors, it doesn't respond to the world and drives into walls.

The robot's Raspberry Pi has many GPIO pins left to add inexpensive sensors for it to interact with its surroundings. This tutorial shows how to use a couple of similar sensor types, put them in useful places, wire them in, and write the code. We'll touch on the trade-offs and limitations, learn how to calibrate the sensors, and make test tracks for line following.

The sensor models chosen are cheap and easily found online, but being IR sensors, they can be dazzled and confused by bright sunlight and some fluorescent lights.

These sensors output digital (on/off) signals, with a dial adjusting the level to go from off to on. They come on carrier boards as 3.3 V compatible modules, making them easier to connect to Raspberry Pi.

You'll Need

> 2 × Obstacle avoidance modules
> **magpi.cc/paTuQW**

> Mini breadboard
> **magpi.cc/CuBDyB**

> 2 × TCRT5000 sensor modules
> **magpi.cc/TtfhiF**

> Jumper wires
> **magpi.cc/pPnpZL**

> Plastic standoffs and screws
> **magpi.cc/Cixtpr**

> 5 mm pitch terminal block
> **magpi.cc/jeThnM**

> Black insulating tape

> Plain white paper

01 Meet the sensors

The two types of sensor in this tutorial both detect reflected infrared light (IR). Obstacle sensors detect objects close enough by a brightness level. Line sensors detect how light/dark the floor is below them.

▲ The robot with sensors part-fitted. See how the cables go through a port in front of the batteries

02 How to use sensors

Obstacle sensors should be front-most, so they don't detect the robot itself. By using two sensors facing forward and slightly to either side, a robot can decide which way it needs to turn to avoid an obstacle.

Line sensors should be under the robot to detect if it has gone off track. A single sensor could sweep across and back over a line, but two sensors, wide enough to go either side of a line, make for a smoother system. They can sense when the line is not in the middle, and which way the robot needs to turn to correct it.

03 Planning and sketching

Test-fit the obstacle sensors on the top of the robot, facing forward and slightly outward.

Sketch the top of the robot showing where to attach the sensors (**Figure 1**). 2.5 mm bolts work for this. Add a 10 mm hole for wires to go through, clear of other features.

Next, test-fit line sensors under the robot, sticking out of the front and about 5 to 7 cm apart.

Finally, sketch the bottom of the robot with line sensors and another wire hole. You may be able to use the threads from the standoffs for Raspberry Pi to hold the line sensors if they are long enough.

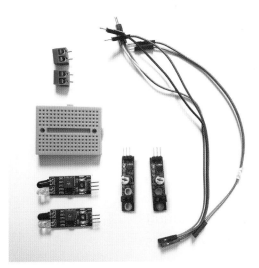

The components needed include terminal blocks on a breadboard, obstacle sensors, line sensors, and jumper wires

04 Drilling holes

Take off the top of the robot, disconnecting the power wires from the motor controller. Use safety gear for all drilling.

Depending on the line sensor placement, you may need to detach the robot's Raspberry Pi and castor before drilling holes in the base of the robot.

Using your sketches, measure out and drill the holes for the sensors. For the cable holes, drill small holes and enlarge them. Remove any excess plastic. Replace your Raspberry Pi and castor, but don't bolt the sensors in or reconnect the batteries just yet.

05 Power distribution

The breadboard gives the sensors access to power rails. 5 mm terminal blocks fit into the breadboard, simplifying connecting the batteries and UBEC. Slot two of the blocks together to make a block of four and pop it into the breadboard, as shown in **Figure 2** (overleaf).

Add breadboard internal wiring connections – pre-cut jumper wires do a tidy job of this. Popping the breadboard lightly into the robot, make the connections to Raspberry Pi 3.3 V and motor board.

The jumper coming from the battery red (positive) wire on the breadboard is a suitable place for an optional power switch.

06 Wiring the sensors

Push a four-way male–female jumper wire through each cable port on the top and bottom for power connections. Use these to wire the sensor VCC/V+ pins to the 3.3 V breadboard row and the GND/G pins to the ground row.

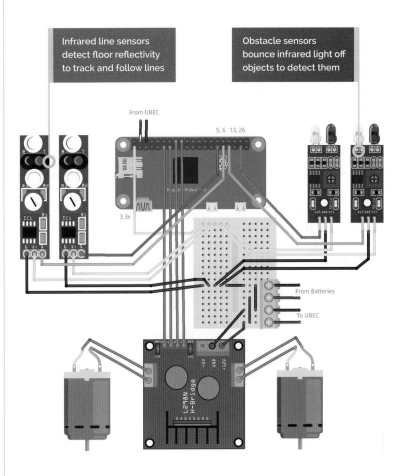

Infrared line sensors detect floor reflectivity to track and follow lines

Obstacle sensors bounce infrared light off objects to detect them

From UBEC

5, 6 13, 26

3.3v

From Batteries

To UBEC

" 5 mm terminal blocks fit into the breadboard, simplifying connecting the batteries and UBEC "

Figure 1

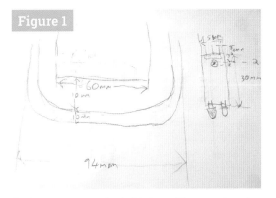

Warning!
Wear goggles

Please use safety goggles and a desk clamp for drilling.

▲ **Figure 1** A start on a sketch of the top and the sensor dimensions. Be prepared to sketch a couple of times, adding more features

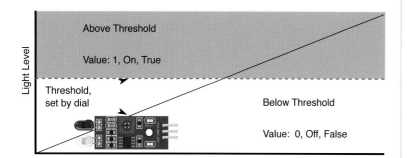

As the light detected goes up and crosses a threshold set by the dial, the sensor will output a logic 1. Below this it's 0

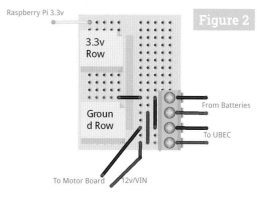

▲ **Figure 2** The power board wiring prepares a power rail for 3.3 V and a rail for GND to connect the sensors to

Sensor digital output pins are labelled D0, DOUT, or S. Wire this pin to a free GPIO pin with a female-female jumper wire. The code uses pins 5 and 6 for the line sensors facing the floor and pins 13 and 26 for the obstacle sensors. Mount the sensors, reassemble the robot, and turn it on.

07 Calibrating the sensors

The sensors may have two LEDs: one is for power and always on, and the other for sensing. Take the robot out of direct sunlight. Ensure nothing is in front of the sensors for a couple of metres.

The sensors have a screwdriver dial on them to adjust the light level they switch on at, which translates to the distance.

Put a reflective obstacle about 10 cm in front of a sensor. Turn the dial slowly to a point the LED turns off. Turn it a tiny way back and it should be on. Wave the obstacle back and forth and watch the LED changing.

08 Sensors in GPIO Zero

GPIO Zero sensors have a value for code to read their state, which a loop could use to set motor values. However, there is a different way to use it.

GPIO Zero has a smart source/value system. A sensor input device can be a 'source' for an output device like a motor, sending a continuous stream of data. So sensor inputs can be virtually wired to affect output, eliminating the loop.

The source/value system has tools to manipulate the data like scaling, negating, or delaying it. Mathematical functions like a sine wave can also be sources. See **magpi.cc/hPmwrv** for more ways to use this.

> ❝ A sensor input device can be a 'source' for an output device like a motor, sending continuous data ❞

09 Driving with sensors

The **obstacle_avoid.py** code makes a `DigitalInputDevice` for each sensor.

Line 10 registers `motor.stop` with the 'atexit' system, guaranteeing when the code stops, for any reason, the motors stop too.

Lines 12 and 13 wire the sensors to the opposite motors, so the robot will turn away from any detected object.

The code scales the sensor values 0 (obstacle detected) and 1 (clear) into motor speed values of –1 (reverse) and 1 (go forward).

obstacle_avoid.py

> Language: **Python 3**

DOWNLOAD THE FULL CODE:
⬇ magpi.cc/QkkmqW

```
001.  from signal import pause
002.  import atexit
003.  import gpiozero
004.  from gpiozero.tools import scaled, negated
005.
006.  robot = gpiozero.Robot(left=(27, 17), right=(24, 23))
007.  left_obstacle_sensor = gpiozero.DigitalInputDevice(13)
008.  right_obstacle_sensor = gpiozero.DigitalInputDevice(26)
009.  # Ensure it will stop
010.  atexit.register(robot.stop)
011.
012.  robot.right_motor.source = scaled(
          left_obstacle_sensor, -1, 1)
013.  robot.left_motor.source = scaled(
          right_obstacle_sensor, -1, 1)
014.
015.  pause()
```

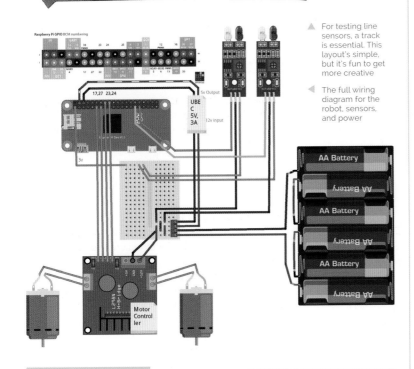

For testing line sensors, a track is essential. This layout's simple, but it's fun to get more creative

The full wiring diagram for the robot, sensors, and power

Run this to see the robot avoid walls and obstacles. The sensors miss obstacles above or below their fields of view, or those too dark and matte to reflect light.

10 Line following

Using sheets of plain white paper (A4 to A1) and black tape (about 20 mm wide), make a single line along the middle of a sheet.

Create a small calibration square of about 40 mm, and put a strip of tape across one end.

Make some curved sections on other paper sheets; keep the turns to less than 45 degrees, and the lines no closer than a robot width apart.

The robot drives forward until the sensor encounters a line. You can use this to make a crossing, by leading tracks to a gap from both horizontal and vertical directions

11 Calibrating line following

Calibrate the line sensors in a similar way to the obstacle sensors. It may be easier to detach the sensor for calibration and reattach it afterwards depending on where the dial is.

Using the calibration square, hold a white area about 2 cm from a line sensor. Slowly turn the sensor dial until the LED changes – wave the paper between black and white to observe the LED changing, and adjust if needed.

You'll need to recalibrate for different lighting or surface conditions.

12 Line following code

The **follow_line.py** code is similar to obstacle avoiding.

Lines 7 and 8 set up GPIO Zero line sensors (based on digital input) on the correct pins.

The line sensors straddle the track. When the sensor detects white (sending 0), it's not crossing the line and so the motor goes forward.

Lines 12 and 13 connect the sensor output so a motor reverses when its sensor crossed the line onto black tape (sending 1). The input source is negated to make 0 the move forward condition.

So that the robot responds to the track before driving past it, scale the source data to go from −0.3 to 0.4.

follow_line.py

> Language: **Python 3**

```
001.  from signal import pause
002.  import atexit
003.  import gpiozero
004.  from gpiozero.tools import scaled, negated
005.
006.  robot = gpiozero.Robot(left=(27, 17), right=(24, 23))
007.  left_line_sensor = gpiozero.LineSensor(5)
008.  right_line_sensor = gpiozero.LineSensor(6)
009.  # Ensure it will stop
010.  atexit.register(robot.stop)
011.
012.  robot.left_motor.source = scaled(negated(left_line_
      sensor), -0.3, 0.4)
013.  robot.right_motor.source = scaled(negated(right_line_
      sensor), -0.3, 0.4)
014.
015.  pause()
```

Make a
Marauder's Map

Make your own Marauder's Map and track your family, pets, and friends (or enemies?) using Bluetooth beacons

MAKER

PJ Evans

PJ is a writer, Raspberry Jammer, and developer. He solemnly swears he is up to no good.

mrpjevans.com

n the Harry Potter series, the Marauder's Map showed you the location of every person in Hogwarts. That map worked with magic, but ours will work with Raspberry Pi and beacons. Bluetooth beacons are low-energy devices that constantly ping out a signal that can be read by any device. Typically, they are used by museums or supermarkets in conjunction with a smartphone app to detect where visitors are and offer relevant information. We're going to flip this and have people carry the beacons with them. Raspberry Pi devices in each room will detect someone's presence and update a web-based map.

01 Get some beacons

Bluetooth Low Energy (BLE) beacons are simple devices that send out a constant signal in the form of a unique code or URL. Bluetooth 4.0 capable devices can detect this signal without needing to pair. There are a few different standards and our project is going to support the two most popular: iBeacon (Apple) and Eddystone (Google). These types of beacons are easily found online and tend to be small button-sized devices capable of running for up to a year on a single battery. Alternatively, software is available so you can use Raspberry Pi boards and microcontrollers (e.g. ESP32) so they act as beacons.

Beacons come in all shapes and sizes. You can even turn Raspberry Pi boards and microcontrollers into beacons

Raspberry Pi Zero W devices are perfect for this project, their on-board Bluetooth detecting the beacons

You'll Need

> At least two rooms

> One Raspberry Pi Zero W per room

> One or more willing participants

> One beacon per participant, e.g. **magpi.cc/iGmnAa**

02 Prepare Raspberry Pi devices

Each person is going to carry a beacon. As they move from room to room, a Raspberry Pi Zero W in each room will detect the presence of the beacon and report it back to a designated server (which can be one of the Raspberry Pi devices). Set up each one with Raspbian Stretch Lite, get them on the WiFi network, then update the software:

```
sudo apt update && sudo apt -y upgrade
```

We need to install a few libraries:

```
sudo apt install python3-pip libbluetooth-dev
sudo pip3 install beacontools[scan]
```

Now we can scan for beacons in Python.

> ❝ Each scanner is going to report its findings back to a central server ❞

03 Get the beacon IDs

Each beacon has either a unique ID code (iBeacon) or broadcasts a web address (Eddystone). If you're using iBeacons and the vendor hasn't supplied the ID, you'll need to discover it by scanning. Create a file containing the code from the **test.py** listing. Save it as **test.py** and run as follows:

```
sudo python3 test.py
```

All being well, you'll see your beacon's transmissions. Make a note of the 32-character string just after 'uuid'. Repeat for each one.

For Eddystones, instructions will be provided on how to set the web address. Set each one to **http://example.org/name**, where 'name' is the person's first name.

04 Install the server software

Each scanner is going to report its findings back to a central server. Pick one of the Raspberry Pi devices for this role, then download the code from **magpi.cc/Hjhtwi** to a directory in your home folder called **beaconmap**. First, install Flask:

```
pip3 install flask
```

The PI-RAUDER'S MAP

Then test the server by running:

```
python3 ~/beaconmap/server/server.py
```

If you're on the same machine, open **http://127.0.0.1:5000** in a browser; otherwise you'll need to replace 127.0.0.1 with the IP address or host name of the server. You'll see a very boring webpage asking 'Where is everyone?'. **CTRL+C** will stop the server.

05 Install the scanners

On each Raspberry Pi, create a file called **scanner.py** and enter the code from the listing here. If you don't fancy typing, install the software package from **magpi.cc/Hjhtwi** and you'll find it in the **scanner** directory.

The code uses Citruz's BeaconTools library to scan the area for beacon signals in bursts of ten seconds. When this happens, the code notes the ID/URL provided and gives it a score of one, with subsequent transmissions incrementing the score. After ten seconds, the scores are sent

▲ Our final web app updates as people move around the house. There's lots of scope for getting creative here

Top Tip

Signal strengths

To avoid all the scanners seeing your beacon, put it on a low power setting if available.

▲ With its small size and smart casing, the Raspberry Pi Zero W makes a tasteful addition to any room

to the server. The server can compare scores for different scanners to work out where someone is, eliminating overlapping areas.

06 Configure and test the server

IDs aren't very useful, so configure the server by editing the **beacons** dictionary. Replace each key with a beacon ID and set the value to {'name': 'name'}. (There are more instructions in the code.) Add as many as you like. When the reports come in from the scanners, we will now know who they are!

You should have something like this:

```
beacons = {
    'b63cc056-6f3a-4a9b-80bf-11ff1c6ff724': {
        'name': 'PJ Evans'
    },
    '144dd069-e22e-418f-b940-c622d64b7252': {
        'name': 'Jazz The Cat'
    }
}
```

Test the server by starting it as before. If you've made any typos in the beacon list, you'll find out now. Leave it running.

Top Tip

Beacon types

There are many different beacon standards; make sure you're using iBeacon or Eddystone for this project.

07 Configure and start your scanners

Edit **scanner.py** and replace the value of `serverUrl` with the address of your server. Make sure '/readings' remains at the end. Then edit **room** to be the Raspberry Pi's room or location name.

On each scanning Raspberry Pi, enter the following command:

```
sudo python3 ~/beaconmap/scanner/scanner.py
```

Each scanner will scan for ten seconds and then report scores back to the server. Check that traffic is flowing correctly. The server will echo incoming data on-screen. If things are not working, check if the server is running a firewall; it needs to allow traffic in through port 5000.

08 Get tracking

With **scanner.py** running on each Raspberry Pi and the server up, take one of your beacons and place it next to one of the scanning Raspberry Pi devices. After ten seconds, refresh the webpage. Your beacon should have been matched and the location displayed. Repeat with another Raspberry Pi. Check the page again. Did it move?

If you get confusing results, the Raspberry Pi boards may be too close to each other (Bluetooth goes through walls!), causing an overlap. If you can set the beacon's power output, it should be as low as possible; this increases battery life and accuracy.

09 Get creative

So far, so good, but it's all a bit boring on the webpage. Grab your crayons and design a map. We're not using any kind of positioning technology, so you can have fun and not worry about accuracy. Make sure it's nice and big as an image (try around 1000×1000 pixels) and save as **beaconmaps/servermap/static/rooms.png**.

You'll also need some avatars, so grab a selfie or screenshot your favourite meme, and create a square image 75×75 for each person and save in the same directory in the format **name.jpg** so it matches the names in the server configuration.

10 Run the advanced server

We've provided a fancier web server using Flask, to make the map a bit more fun. Stop the standard server (**CTRL+C**), configure the **server.py** file in **servermap** as before, then run the following command:

```
python3 beaconmap/servermap/server.py
```

Give the scanners time to report in and have a look at the page. It will probably be a bit of a mess, but you can have a look at the code for instructions on how to adjust things so your avatars appear in the right places. Make sure everyone is appearing in the correct location, then stop the server and scanners.

> ## " Make sure everything starts on boot, then runs in the background "

11 Automate all the things
The final step is to make sure everything starts on boot, then runs in the background so you don't need to have a Terminal open all the time. There are many methods, but the easiest way to do this is to edit the **rc.local** file and add our requirements.

```
sudo nano /etc/rc.local
```

Before the last line that reads **exit 0**, insert a new line and add the following lines.
For the server:

```
/usr/bin/python3 /home/pi/beaconmap/servermap/
server.py &
```

For each scanner:

```
/usr/bin/python3 /home/pi/beaconmap/scanner/
scanner.py &
```

To test, reboot each device. Everything should now run in the background.

12 More with beacons
Now let your imagination get involved. What else can you do with these beacons? Try modifying the server to alert your smartphone when someone arrives in a certain place. Or how about a digital Easter egg hunt? Give everyone a battery-powered Raspberry Pi Zero W and hide beacons. Your score could be generated automatically as you find them. Could you make a box that only opens when someone carrying the right beacon approaches? Over to you. 🔧

scanner.py

> Language: **Python 3**

DOWNLOAD THE FULL CODE:
⊙ **magpi.cc/Hjhtwi**

```python
001.  import time
002.  import requests
003.  from beacontools import BeaconScanner
004.
005.  serverUrl = "http://127.0.0.1:5000/readings"
006.  room = "Kitchen"
007.  beacons = {}
008.
009.
010.  # This function is called whenever a packet is detected
011.  def callback(bt_addr, rssi, packet, additional_info):
012.
013.      # Parse out the type of beacon
014.      typeOfBeacon = type(packet).__name__.split(".").pop()
015.
016.      # Get the ID of the beacon
017.      if typeOfBeacon == "EddystoneURLFrame":
018.          beaconId = packet.url
019.      elif typeOfBeacon == "IBeaconAdvertisement":
020.          beaconId = packet.uuid
021.
022.      # Track how many times we've seen this beacon
023.      if beaconId not in beacons:
024.          beacons[beaconId] = 1
025.      else:
026.          beacons[beaconId] += 1
027.
028.  # Scan for all advertisements from beacons
029.  print('Starting beacon scanner')
030.  scanner = BeaconScanner(callback)
031.  scanner.start()
032.
033.  while True:
034.
035.      # Allow a 10-second sample to come through
036.      print('Waiting 10 seconds')
037.      time.sleep(10)
038.
039.      # Now send the current scores to the server
040.      print('Sending to server')
041.      try:
042.          response = requests.put(serverUrl, json={"room": room,
043.                                  "beacons": beacons})
044.          if response.status_code == 200:
045.              print('Success')
046.          else:
047.              print('Got response code: ' + str(response.status_code))
048.      except:
049.          print("Communication error")
050.
051.      # Clean the scores
052.      beacons = {}
```

Use AI to build
a plant monitor

Measure the height and width of objects with your Raspberry Pi
using only the Raspberry Pi Camera Module and OpenCV

MAKER

**PJ
Evans**

PJ is a writer,
developer, and
Milton Keynes
Jam wrangler. He
has terrible taste
in movies.

mrpjevans.com

Do you know how tall your plant is? Do you
wonder how high it is growing? Does using
a ruler just sound like effort? Well, we have
the tutorial for you. We're going to measure a
plant just using images taken with a Raspberry Pi
Camera Module. In the process, we'll introduce
you to OpenCV, a powerful tool for image analysis
and object recognition. By comparing your plant
to a static object, OpenCV can be used to estimate
its current height, all without touching. By adding
data from other sensors, such as temperature or
humidity, you too can build a smart plant.

01 Pick a spot

We're going to be using OpenCV (version 2)
to measure our plant. To make sure our
measurement is as accurate as possible, several

factors need to be considered. The background
must be plain and contrast against the plant;
black or white is ideal. The area you are using
must be very well lit when the image is taken.
Finally, the camera must be 90 degrees to the
plant, facing it directly. Any angles on any axis
will cause poor measurement. If you are mounting
the camera at a significant distance, you may want
to consider a telephoto lens.

02 Get a plant and object

So how do we pull off this measuring
trick? We can estimate the height of an object by
comparing it with another object that is a fixed
and known width and height. The image taken
by the camera needs to include both objects. Get
a rectangular object of a comparable size to your

You'll Need

> Pi Camera Module
magpi.cc/camera

> A houseplant

> Ruler or similar
rectangular object

> Contrasting, plain
backdrop

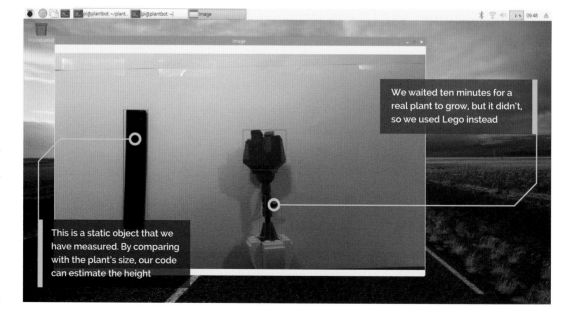

We waited ten minutes for a
real plant to grow, but it didn't,
so we used Lego instead

This is a static object that we
have measured. By comparing
with the plant's size, our code
can estimate the height

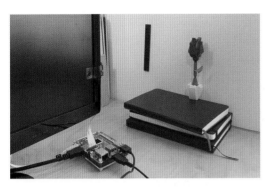

Our testing rig. Note the camera is as close to 90 degrees to the plant as we could manage

plant; we used a long piece of Lego. A metal ruler would also work well. Mount your object in front of the camera so it is on the left of the image and the same distance from the camera as your plant.

03 Prepare your Raspberry Pi

The software used to analyse the image is the powerful OpenCV library and its Python bindings. You can use any current Raspberry Pi for this project, but the higher-end 3 or 4 will be much quicker at processing the image. OpenCV requires an X Window system in place, so we need to start with Raspbian Stretch including the Raspberry Pi Desktop. Once ready to go and on the network, make sure everything is up-to-date by running:

```
sudo apt update && sudo apt -y upgrade
```

Finally, install the camera, if you haven't already, and enable it in Preferences > Raspberry Pi Configuration > Interfaces, then reboot.

04 Install dependencies

OpenCV is a bit trickier to install than most packages. It has lots of dependencies (additional software it 'depends' on) that are not installed alongside it. We also need a few other things to get OpenCV talking to the camera, so open up a Terminal window and run the following commands:

```
sudo apt install python3-pip libatlas-base-
dev libhdf5-100 libjasper1 libqtcore4 libqt4-
test libqtgui4

pip3 install imutils opencv-contrib-python
picamera[array]
```

plantbot.py

DOWNLOAD THE FULL CODE:
magpi.cc/omacNA

> Language: **Python 3**

```python
001. import argparse
002. import imutils.contours
003. import cv2
004. from picamera.array import PiRGBArray
005. from picamera import PiCamera
006. from time import sleep
007.
008. # Get our options
009. parser = argparse.ArgumentParser(description='Object
     height measurement')
010. parser.add_argument("-w", "--width", type=float,
     required=True,
011.                     help=
     "width of the left-most object in the image")
012. args = vars(parser.parse_args())
013.
014. # Take a photo
015. camera = PiCamera()
016. rawCapture = PiRGBArray(camera)
017. sleep(0.1)
018. camera.capture(rawCapture, format="bgr")
019. image = rawCapture.array
020.
021. # Cover to grayscale and blur
022. greyscale = cv2.cvtColor(image, cv2.COLOR_BGR2GRAY)
023. greyscale = cv2.GaussianBlur(greyscale, (7, 7), 0)
024.
025. # Detect edges and close gaps
026. canny_output = cv2.Canny(greyscale, 50, 100)
027. canny_output = cv2.dilate(canny_output, None,
     iterations=1)
028. canny_output = cv2.erode(canny_output, None, iterations=1)
029.
030. # Get the contours of the shapes, sort l-to-r and create
     boxes
031. _, contours, _ = cv2.findContours(canny_output, cv2.RETR_
     EXTERNAL,
032.                                   cv2.CHAIN_APPROX_SIMPLE)
033. if len(contours) < 2:
034.     print("Couldn't detect two or more objects")
035.     exit(0)
036.
037. (contours, _) = imutils.contours.sort_contours(contours)
038. contours_poly = [None]*len(contours)
```

Although piwheels speeds things up considerably, you can expect the second command to take a little while to run, so now's a good time to get a cup of tea.

05 Calibration test

Before we start running code, make sure everything is lined up as you would like it. The easiest way to do this is to open a Terminal prompt on the Desktop and run the following command:

```
raspistill -t 0
```

This puts the camera in 'preview' mode (you'll need to be directly connected to see this – it won't work over VNC without extra configuration). Study the image carefully. Is the camera at an angle? Is it too high or low? Is there enough contrast between your object, plant, and background? How's the light? Press **CTRL+C** when you're finished.

06 Coding time

Enter the code from the **plantbot.py** listing in your editor of choice and save it. If you don't fancy typing, you can download the code and some sample images from **magpi.cc/akmZsW**. The code takes an image from the Raspberry Pi Camera as a stream and sends it to OpenCV. The image is then placed through some cleaning filters and the shapes detected. We know the nearest shape to the 0,0 'origin' (the top-left of the image) will be our calibration object. So long as we know the width of that object, we can estimate the size of others.

▼ This overlay shows how OpenCV has mapped the test image. It can be far more accurate, but this faster method meets our needs

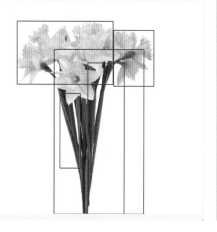

07 Run the code

Let's take our first image. Carefully measure the width of your calibration object. Ours was 16 mm. To run the code, enter the following command in the same location as your **plantbot.py** file:

```
python3 plantbot.py -w 16
```

Replace the '16' with the width of your calibration object. After a few seconds, you'll see the image with, hopefully, a light blue rectangle around the leftmost object and one or more boxes around the plant. As the plant is an irregular object, OpenCV will find lots of 'bounding boxes.' The code takes the highest and lowest to calculate the height.

08 Big plant data

Once you've had fun measuring all the things, you can alter the code to write the results to a file. Add the following to the 'import' section at the start of the code:

```
import datetime
```

Now remove the final six lines (the first one starts 'print') and replace with:

```
print("\"" + str(datetime.datetime.now()) +
"\"," + str(plantHeight))
```

This version can be employed to create a CSV file of measurements that may be used to create graphs so you can track your plant's growth.

09 Keep on schedule

Let's say you wanted to take a measurement every day. You can do this by using Raspbian's cron system. Cron is an easy way to run scripts on a regular basis. To access the main cron file (known as a crontab), run the following:

```
sudo nano /etc/crontab
```

At the bottom of the file, add this line:

```
0 14    * * *   pi      python3 /home/
pi/plantbot/plantbot.py -w 16 >> /home/pi/
plantbot.csv
```

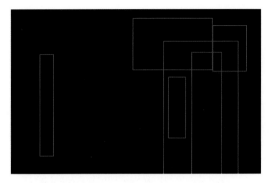

An example of what OpenCV sees, with the calibration object on the left, and some daffodils on the right

This tells cron to run our script every day at 2pm (that's the '0 14' bit) and append its output to **plantbot.csv**, which can be read by any popular spreadsheet app.

10 Got water?

Now you're collecting data on your beloved plant's height, why not add other data too? A common Raspberry Pi project is to use soil moisture sensors. These are inexpensive and widely available (**magpi.cc/seyhTc**). See if you can change the script to get moisture data and include it in the CSV output. You could even send alerts when your plant needs some water. If you imported weather data from free public APIs such as **openweathermap.org/api** as well, you could see how changing conditions are affecting your plant!

11 More than plants

You can measure anything you want with this project, even growing youngsters, but why stop there? You've now had a taste of the power of OpenCV, which has many more capabilities than we've covered here. It is a powerhouse of computer vision that includes machine-learning capabilities that allow you to train your Raspberry Pi to recognise objects, faces, and more. It's a popular choice for robot builders and plays a major part in the autonomous challenges of Pi Wars. We can especially recommend **pyimagesearch.com**, which provided inspiration for this tutorial.

Top Tip

No plant?

If you don't have a plant or suitable environment, we've provided some sample images here: **magpi.cc/omacNA**

> You can measure anything you want with this project, even growing youngsters, but why stop there?

plantbot.py (continued)

> Language: **Python 3**

```python
039. boundRect = [None]*len(contours)
040. for i, c in enumerate(contours):
041.     contours_poly[i] = cv2.approxPolyDP(c, 3, True)
042.     boundRect[i] = cv2.boundingRect(contours_poly[i])
043.
044. output_image = image.copy()
045. mmPerPixel = args["width"] / boundRect[0][2]
046. highestRect = 1000
047. lowestRect = 0
048.
049. for i in range(1, len(contours)):
050.
051.     # Too smol?
052.     if boundRect[i][2] < 50 or boundRect[i][3] < 50:
053.         continue
054.
055.     # The first rectangle is our control, so set the ratio
056.     if highestRect > boundRect[i][1]:
057.         highestRect = boundRect[i][1]
058.     if lowestRect < (boundRect[i][1] + boundRect[i][3]):
059.         lowestRect = (boundRect[i][1] + boundRect[i][3])
060.
061.     # Create a boundary box
062.     cv2.rectangle(output_image, (int(boundRect[i][0]),
     int(boundRect[i][1])),
063.                 (int(boundRect[i][0] + boundRect[i][2]),
064.                 int(boundRect[i][1] + boundRect[i][3])),
     (255, 0, 0), 2)
065.
066. # Calculate the size of our plant
067. plantHeight = (lowestRect - highestRect) * mmPerPixel
068. print("Plant height is {0:.0f}mm".format(plantHeight))
069.
070. # Resize and display the image (press key to exit)
071. resized_image = cv2.resize(output_image, (1280, 720))
072. cv2.imshow("Image", resized_image)
073. cv2.waitKey(0)
```

Make comics
from TV recordings

Convert a recording from the Raspberry Pi TV HAT into a comic book
and read the latest Doctor Who episode on your Kindle

MAKER

PJ
Evans

PJ is a writer,
developer, and
Milton Keynes
Jam wrangler. He
has terrible taste
in movies.

mrpjevans.com

Read any good telly recently? How about
catching up on your favourites shows on
an e-ink reader or tablet? Sounds silly but
it can be a nice, peaceful alternative to sit back
and flick through *Holby City*. OK, we're reaching
a little bit. The real fun here is learning about
video and image manipulation, optical character
recognition, generating PDFs in code, and
using Python as a powerful scripting language
to pull several tools together. We'll take the
raw recording produced by the Raspberry Pi TV
HAT and create a PDF document, complete with
captions taken from subtitles.

01 Get recording

Before starting, make sure you have your
Raspberry Pi set up with a TV HAT and Tvheadend
installed (see 'You'll Need' box for a helpful link).
You will need a recording from Tvheadend (it
doesn't matter what, but maybe the news wouldn't
be the most exciting choice). You can select any
programme and record it, then find the recording

▼ The TV HAT connects
to a digital antenna,
giving your Raspberry
Pi over-the-air
access to over 80 TV
channels and radio

under 'Digital Video Recorder' then 'Finished
Recordings'. From here you can download the
file or you can find recordings in **/var/lib/hts**.
Tvheadend records in the original broadcast
MPEG-2 TS format (or 'transport stream').

02 Install dependencies

The process of converting a recording to
a PDF is going to take several discrete stages.
These include video extraction, optical character
recognition (OCR), and generating PDFs. Not all
of this is easily within Python's reach, so we'll
use Python to manage the process, delegating the
'heavy lifting' to some command-line utilities.
Their purposes will become apparent as we go
through the tutorial. Here's what you need to do at
the command-line:

```
sudo apt update && sudo apt -y upgrade
sudo apt install git python3-pip ffmpeg
imagemagick
pip3 install fpdf arrow
```

03 Compile and install ccextractor

The utility 'ccextractor' is able to remove
subtitles from DVB (Digital Video Broadcasting)
recordings. Unfortunately, this application is not
available in the APT repositories, so we're going to
have to compile it ourselves. We'll use Git, which
we installed in the previous step, to download the
source code from its repository. Then we'll install
its dependencies (other programs it relies on)
before compiling and installing the app.

```
cd
git clone https://github.com/CCExtractor/
```

DOWNLOAD THE FULL CODE:
magpi.cc/swtRoH

Stills are extracted from the source video file based on subtitle times and scene changes

Subtitles are extracted, converted to text, and then rendered in a comic book-style font

```
ccextractor.git
  sudo apt install -y libglfw3-dev cmake
gcc libcurl4-gnutls-dev tesseract-ocr
tesseract-ocr-dev libleptonica-dev
  cd ccextractor/linux
  ./build
  sudo mv ./ccextractor /usr/local/bin/
```

04 Install the script

As this is a series of steps potentially involving hundreds if not thousands of files, we've provided a Python script to control the process. It's a bit on the large side to type in manually, so again we'll use Git. To get the code on to your Raspberry Pi, enter the following commands:

```
cd
  git clone https://github.com/mrpjevans/
comical.git
```

You will now have a new directory, **comical**, containing the script and a few other files we need.

05 Extract the subtitles

Rather than just run the entire script, which wouldn't show us much, let's run it in

> **Before starting, make sure you have your Raspberry Pi set up with a TV HAT and Tvheadend installed**

stages. Make sure you know the path to your recording. We're using the public domain movie *Plan 9 From Outer Space*, regarded as one of the worst films ever made. The first job is to extract the subtitles from the video so we can process them and use them as captions.

```
cd ~/comical
python3 comical.py -i plan9.ts --extract
```

A folder, **plan9.d**, is created, containing a PNG image file for each subtitle. An XML file, **plan9.xml**, contains the timing information for each title.

06 Cleaning up

So why are our subtitles images? It's because that's the European digital broadcast standard. Subtitles in DVB are actually a second video stream. To make use of them, we'll need to take the PNGs that ccextractor created and perform

You'll Need

> Raspberry Pi TV HAT
  magpi.cc/oBXuot

> Tvheadend installation
  magpi.cc/QCkFdt

> e-book reader or tablet

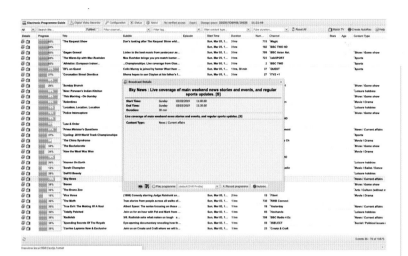

Strange name and strange interface, but Tvheadend more than makes up for that with its seemingly endless features

optical character recognition on them. Currently they're too small to be recognised accurately by the OCR application Tesseract. So, we'll use the ImageMagick utility 'mogrify' to resize them and greyscale them.

```
python3 comical.py -i plan9.ts --clean
```

If you have a look in the directory, you'll see the subtitles are now large and monochrome.

07 OCR With Tesseract

Tesseract is a remarkable utility originally developed by Hewlett-Packard and open-sourced. Now you have it installed on your Raspberry Pi, you can use it for many other purposes. To convert something into text, just run:

```
tesseract <image file> <output file>
```

Our script reads in every image in the directory and sends it to Tesseract for processing. At this size, you can expect a good level of accuracy from DVB titles.

```
python3 comical.py -i plan9.ts --ocr
```

In the same directory you'll now see a matching '.txt' file for each graphic subtitle.

Top Tip 👍

Not just DVB

The script will also work with other video formats that have a supported video subtitle track.

08 Extract images

The next part of our script will extract a single still image for each subtitle based on that subtitle's timestamp. To get the timestamps, we

use Python's built-in XML parsing libraries. For each timestamp, we then ask ffmpeg (a Swiss Army knife for video processing) to extract a still image as a JPEG and save it in (in our case) a new directory called **plan9_process**. The file name represents the time code at which it appears. We also copy across the subtitle with a matching file name.

```
python3 comical.py -i plan9.ts --images
```

09 A picture worth a thousand words

Have a look in your equivalent of **plan9_process**. We've got everything we need to build our PDF. Right? Well, yes, provided there's no break in dialogue, which seems unlikely. What about scenes with no subtitles? Again, ffmpeg comes to our rescue. An advanced filter can detect when a significant amount of the screen changes, denoting a scene change. Our script will ask ffmpeg to detect every scene change and then extract further JPEG images, ignoring any that are within a second of a subtitle image.

```
python3 comical.py -i plan9.ts
--detectscenes
python3 comical.py -i plan9.ts
--extractscenes
```

Your **_process** directory is now populated with the additional images.

10 Build it!

The final part of the script will take all the images and text files and convert them into a PDF for you to enjoy.

```
python3 comical.py -i plan9.ts -o plan9.
pdf --build
```

This part of the script uses the fpdf Python library to lay out each image in a 2×3 grid, adding pages as needed. Where there is a matching subtitle, it is placed below the image. To give the final result a bit more of a graphic novel feel, there is a comic book-style font included in the comical repository which is used by fpdf when rendering text.

> Tesseract is a remarkable utility originally developed by Hewlett-Packard and open-sourced "

11 Adjusting fonts

You might find that sometimes, dependent on the subtitle lengths, the captions can overflow; or that the font size isn't large enough, with too much white space. The script provides a few arguments that can be specified on the command line to help with this:

```
python3 comical.py -i plan9.ts -o plan9.
pdf --build --fontsize 8 --lineheight 5
--offset 68
```

Here, `fontsize` sets the size of the font. This needs to be in step with `lineheight`, which sets the vertical spacing between lines. `offset` sets the position of the first line of text below the image. The default settings are shown above.

12 Automating and fine-tuning

The **comical.py** script comes with a number of arguments to control its behaviour. In the tutorial we've gone step by step, but you could have just run the following:

```
python3 comical.py -i plan9.ts -o plan9.
pdf --full
```

This performs every step in one operation. You can also do a pre-build:

```
python3 comical.py -i plan9.ts -o plan9.
pdf --prebuild
```

This performs every step except building the PDF, as you might want to remove unwanted images and subtitles to crop the PDF to the things in which you are interested. Delete the unwanted files from your **_process** directory and then run:

```
python3 comical.py -i plan9.ts -o plan9.
pdf --build
```

13 Make it your own

You could regard this project as a bit frivolous, but in the process of putting it together we've looked at several cool technologies such as video manipulation and optical character recognition. Examine the script code to see how we use Python to link all these different utilities together and marshal the data flowing between them. Why not see if you can improve on the results? Some ideas include adding filters to the images to give a graphic novel appearance, watching the recordings folder to trigger automatic conversion, creating glitch art, or mashing up different recordings.

And remember my friend, future events such as these will affect you in the future.

And remember my friend, future events such as these will affect you in the future.

And remember my friend, future events such as these will affect you in the future.

◀ This is an original subtitle. Yes, the script is really that bad. It's a PNG taken from the subtitle video stream

◀ We use ImageMagick's mogrify utility to remove colour, invert the image, and increase size by 400% to improve OCR accuracy.

◀ Tesseract reads in the image and produces text output. As you can see, it's very accurate if the source material is clear

AND REMEMBER MY FRIEND, FUTURE EVENTS SUCH AS THESE WILL AFFECT YOU IN THE FUTURE.

▲ The final step is to produce a more fun caption for the panel by rendering it in a comic book font

MAKER

PJ
Evans

PJ is a writer, trainer, and freelance software engineer. His son can't 'borrow' the car as easily any more.

mrpjevans.com

Car Spy Pi

Who's that parked on the driveway?
Find out automatically using ANPR

Automatic number-plate recognition (ANPR) is becoming more and more commonplace. Once the exclusive realm of the police, the technology used to accurately read car number-plates can now be found in supermarket and airport car parks. It wasn't long ago that this technology was extremely expensive to purchase and implement. Now, even the Raspberry Pi has the ability to read number-plates with high accuracy using the Raspberry Pi Camera Module and open-source software. Let's demonstrate what's possible by building a system to detect and alert when a car comes onto the driveway.

01 Pick a spot

First things first: where are we going to put it? Although this project has lots of applications, we're going to see who's home (or not) by reading number-plates of cars coming and going on a driveway. This means the Raspberry Pi is probably going to live outside; therefore, many environmental constraints come into place. You'll need USB 5 V power to Raspberry Pi and a mounting position suitable for reading plates, although the software is surprisingly tolerant of angles and heights, so don't worry too much if you can't get it perfectly aligned.

02 Get an enclosure

As our Raspberry Pi will live outside (unless you have a well-placed window), you'll need an appropriate enclosure. For a proper build, get an IP67 waterproof case (e.g. **magpi.cc/epzGcX**). We're opting for homemade, and are using a Raspberry Pi 3 A+ with the RainBerry – a 3D-printable case that, once you add some rubber seals, provides adequate protection. Make sure whatever you choose has a hole for the camera.

You'll Need

> Pi Camera Module
magpi.cc/camera

> Suitable outdoor enclosure, e.g.
magpi.cc/hOVWBP

> Pushover account (optional)
pushover.net

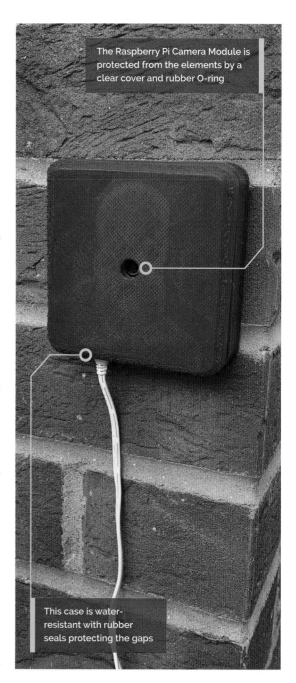

The Raspberry Pi Camera Module is protected from the elements by a clear cover and rubber O-ring

This case is water-resistant with rubber seals protecting the gaps

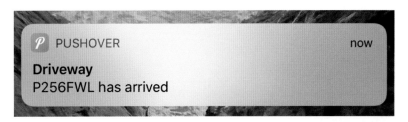

PUSHOVER   now

Driveway
P256FWL has arrived

03 Prepare your Raspberry Pi

As we don't need a graphical user interface, Raspbian Stretch Lite is our operating system of choice. Any Raspberry Pi can handle this task, although if you want the fastest image processing possible, you probably want to avoid the Zero models and get a nice, zippy Raspberry Pi 3 or 4. Get your operating system set up, make sure you've done the necessary `sudo apt update && sudo apt -y upgrade` and have configured WiFi if you're not using Ethernet. Finally, make sure your Raspberry Pi Camera Module is connected and enabled. You can check this by running `sudo raspi-config` and looking under 'Interfacing Options'.

04 Install openALPR

Thankfully, we don't need to be experts in machine learning and image processing to implement ANPR. An open-source project, openALPR provides fast and accurate processing just from a camera image. 'ALPR' is not a mistake: this US project is 'Automatic *License* Plate Recognition'. Thanks to APT, installation is straightforward. At the command line, enter the following:

```
sudo apt install openalpr openalpr-daemon openalpr-utils libopenalpr-dev
```

This may take a while, as many supporting packages need to be installed, such as Tesseract, an open-source optical character recognition (OCR) tool. This, coupled with code that identifies a number-plate, is what works the magic.

▼ Protect your Raspberry Pi with a waterproof case

▲ When a car arrives or leaves our driveway, we receive an alert in seconds

05 Time for a test

Once installed, you'll be unceremoniously dropped back to the command prompt. OpenALPR has installed a command-line tool to make testing its capabilities easier and they have also kindly provided a sample image. On the command line, enter the following:

```
cd
wget http://plates.openalpr.com/ea7the.jpg
```

This is a sample USA plate image and a tough one too. Wget grabs the file from the web and places it in your home directory. Let's see if we can recognise it:

```
alpr -c us ea7the.jpg
```

All being well, you'll see a report on screen. The most likely 'match' should be the same as the file name: EA7THE.

06 Install Python libraries

We can use openALPR in Python, too. First, install the libraries using pip. If you don't have pip installed, run:

```
sudo apt install python-pip
```

To install the libraries:

```
pip install openalpr picamera python-pushover
```

Now test everything is working by running Python and enter the following code line by line at the >>> prompt:

```
import json
from openalpr import Alpr

alpr = Alpr("us", "/etc/openalpr/openalpr.conf", "/usr/share/openalpr/runtime_data")
results = alpr.recognize_file("/home/pi/ea7the.jpg")
print(json.dumps(results, indent=4))
alpr.unload()
```

▲ Our target. The software does a great job of recognising number-plates from different heights and angles

is found, we get the number. If there has been a change, an alert is sent to Pushover, which is then forwarded to any registered mobile devices.

If you've not seen JSON-formatted text before, this might seem a bit much, but you should see the correct plate number returned as the first result.

07 Get a Pushover token

So that we can get an alert when a car arrives or leaves, we're using old favourite Pushover (**pushover.net**), which makes sending notifications to mobile phones a breeze. There's a free trial, after which it's a flat fee of $4.99 per device, with no subscription or limits. Once logged in, go to 'Your Applications' and make a note of your User Key. Then click 'Create an Application/API Token'. Call it 'ANPR', leave all the other fields blank, and click 'Create Application'. Now make a note of the API Token; you'll need this and the User Key for your code.

08 Typing time

Now you have everything you need to create your ANPR application. Enter the code listing shown here or download it from **magpi.cc/VEsaCg**. Save it as **anpr.py** in your home directory. Edit the file and enter your User and App tokens where prompted. Save the file, then test by entering:

```
python anpr.py
```

The code makes use of the Raspberry Pi Camera Module and openALPR in tandem. Every five seconds, the camera takes a picture which is passed to openALPR for analysis. If a licence plate

09 Make your list

If you want to have more friendly names rather than just the plate number, try adding a Python dictionary just after the import statements, like this:

```
lookup = {
            "ABC123": "Steve McQueen",
            "ZXY123": "Lewis Hamilton"
          }
```

Then change all instances of `number_plate` in the alert text as follows:

```
lookup[number_plate]
```

Now you'll get a friendly name instead. See if you can handle what happens if the plate isn't recognised.

10 Run on boot

A key part of any 'hands-free' Raspberry Pi installation is ensuring that in the event of a power failure, the required services start up again. There are many ways of doing this; we're going use one of the simpler methods.

```
sudo nano /etc/rc.local
```

Find the final line, `exit 0` and enter the following on the line above:

```
#Start ANPR Monitoring
/usr/bin/python /home/pi/anpr.py
```

Press **CTRL+X** then **Y** to save the file. Finally, run the earlier pip command again, using sudo this time to install the libraries for the root user:

```
sudo pip install openalpr picamera
python-pushover
```

On reboot, the code will start up and run in the background.

anpr.py

> Language: **Python 3**

```python
001.  from openalpr import Alpr
002.  from picamera import PiCamera
003.  from time import sleep
004.  import pushover
005.
006.  # Pushover settings
007.  PUSHOVER_USER_KEY = "<REPLACE WITH USER KEY>"
008.  PUSHOVER_APP_TOKEN = "<REPLACE WITH APP TOKEN>"
009.
010.  # 'gb' means we want to recognise UK plates, many
      others are available
011.  alpr = Alpr("gb", "/etc/openalpr/openalpr.conf",
012.              "/usr/share/openalpr/runtime_data")
013.  camera = PiCamera()
014.  pushover.init(PUSHOVER_APP_TOKEN)
015.  last_seen = None
016.
017.  try:
018.      # Let's loop forever:
019.      while True:
020.
021.          # Take a photo
022.          print('Taking a photo')
023.          camera.capture('/home/pi/latest.jpg')
024.
025.          # Ask OpenALPR what it thinks
026.          analysis = alpr.recognize_file(
      "/home/pi/latest.jpg")
027.
028.          # If no results, no car!
029.          if len(analysis['results']) == 0:
030.              print('No number plate detected')
031.
032.              # Has a car left?
033.              if last_seen is not None:
034.                  pushover.Client(
      PUSHOVER_USER_KEY).send_message(
035.                      last_seen + " left",
036.                      title="Driveway")
037.
038.                  last_seen = None
039.
040.          else:
041.              number_plate =
      analysis['results'][0]['plate']
042.              print('Number plate detected: ' +
      number_plate)
043.
044.              # Has there been a change?
045.              if last_seen is None:
046.                  pushover.Client(
      PUSHOVER_USER_KEY).send_message(
047.                      number_plate + " has arrived",
      title="Driveway")
048.              elif number_plate != last_seen:
049.                  pushover.Client(
      PUSHOVER_USER_KEY).send_message(
050.                      number_plate + " arrived  and "
      + last_seen + " left",
051.                      title="Driveway")
052.
053.              last_seen = number_plate
054.
055.          # Wait for five seconds
056.          sleep(5)
057.
058.  except KeyboardInterrupt:
059.      print('Shutting down')
060.      alpr.unload()
061.
```

11 Add logging and curfews

One use of this installation is to track the times cars come and go. This can be especially useful for young drivers who have curfew restrictions on their insurance. See if you can augment the code to check whether a registration plate has not been seen after a certain time. For example, if your younger family members have such a restriction, send them an alert to their phone if their car isn't in the driveway 30 minutes beforehand. You might save them an insurance premium increase! Also, why not log all the comings and goings to a file? A bit of data analysis might help reduce car usage or fuel costs.

12 Make it your own

As ever, it's over to you. Now you have the ability to track and record registration plates, there are many different applications for you to explore. Since all the analysis of the image is done locally, no internet connection is required for the system to work. Is there someone 'borrowing' your parking space at work? Catch 'em in the act! Why not take your Car Spy Pi on the road? It could record every vehicle you encounter, which may be useful should something untoward happen. Combine a Raspberry Pi Zero with a ZeroView (**magpi.cc/eBnYrZ**) and you're all set. 🔲

Reviews

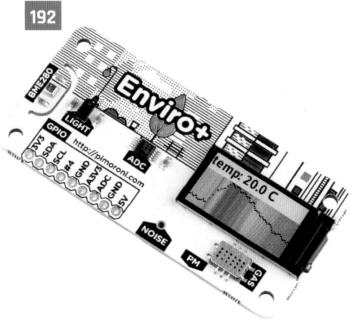

Reviews

Pibow Coupé 4
and Fan SHIM

PIBOW COUPÉ 4 ▶ Pimoroni ▶ **magpi.cc/enLWLt** ▶ £9/$9 | **FAN SHIM** ▶ Pimoroni ▶ **magpi.cc/qZYBWd** ▶ £10/$10

SPECS

COLOURS:
Rainbow, Red,
Ninja

WEIGHT:
50 g

DIMENSIONS:
99 × 66 × 15 mm

Raspberry Pi 4's layout means new cases, including a new version of the famous Pibow. **Rob Zwetsloot** checks it out, along with a cooling Fan SHIM

One glance at Raspberry Pi 4 and you'll notice that it looks quite a bit different from your standard Raspberry Pi Model B. Since the original Raspberry Pi B+ came out, all Raspberry Pi boards have had a standard port layout, but with new tech comes new features, which is why Raspberry Pi 4 has some extra ports and a shuffling around of the USB and Ethernet.

You can read why the board layout has changed in the feature starting on page 8; however, it does mean you'll either need to hack apart an old Raspberry Pi case or get a new one. Which is where Pimoroni comes in, as it so often does, with the new Pibow Coupé for Raspberry Pi 4.

Currently the Coupé is the only case style that Pimoroni is making for Raspberry Pi 4, lacking the top layers of a full Pibow. However, it still partially

▲ The Fan SHIM is very small but pretty powerful

covers and protects a Raspberry Pi 4 while giving full access to the GPIO and ports. It also comes with a bonus feature: the ability to add a special fan to help keep your Raspberry Pi 4 nice and cool.

Treasure box
Construction of the case is simple. There are five numbered slices of plastic that slot on and around Raspberry Pi 4, which are then tightened with some plastic nuts and bolts. We were able to put it together in a couple of minutes; if you're having trouble, however, there's a handy online build video from Pimoroni.

If you've ever had a Pibow case, you know the score – it's sturdy, light, and looks pretty nice. Importantly, it keeps a Raspberry Pi protected pretty well from dust, grubby fingers, and other environmental hazards. The top plate labels what each input/output port is, which is especially handy with the shuffles and additions for this hardware release.

▲ The latest Pimoroni Pibow Coupé

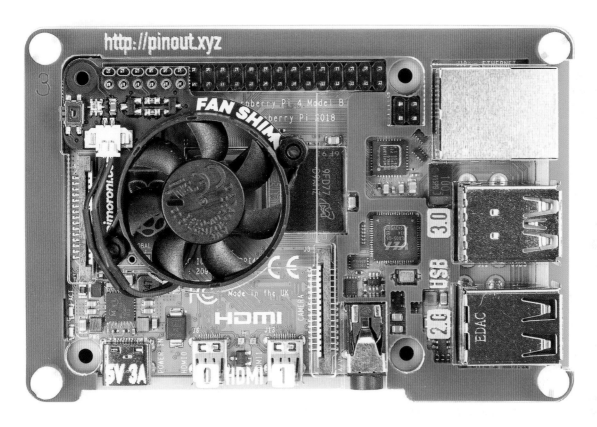

> **Specific care has been taken to keep the important chips open to the air**

Specific care has been taken to keep the important chips open to the air, allowing for better ventilation of the board. While you don't need any special heat sinks or fans for Raspberry Pi 4, it doesn't hurt to add them, and the Pibow Coupé is specifically designed to fit the new Fan SHIM.

Biggest fan

Again, assembly is simple. Screw the fan onto the PCB, plug it in, then you can slip it over the GPIO pins while your Raspberry Pi is off. Quick and easy. However, you will need to install some software to control it, which is where it gets really fun.

Very basically, the software lets you turn the fan on and off. However, with different scripts, you can have it activate at specific temperatures and even change what the LED and button do. It's very hackable and, according to tests, very effective as well. It also doesn't sound like a hoover, which is always a bonus with a case fan.

Raspberry Pi computers have been around for over seven years now, and Pimoroni making a nice case for one is hardly a shocker – however, it's nice to see that the first round of Raspberry Pi 4 cases are great. ▯

▲ The case design allows for easily adding the Fan SHIM

▲ The Fan SHIM is very small but pretty powerful

Verdict

A great first case for Raspberry Pi 4, keeping it very accessible for tinkering and hacking. The optional Fan SHIM is ideal for cooling it during periods of high CPU load.

9/10

Rocky**Borg**

▶ PiBorg ▶ **magpi.cc/dfcqRi** ▶ £99 / $135

SPECS

DIMENSIONS:
16×24×13 cm
(W×L×H)

WHEELS:
70 cm front,
65 cm rear

SPEED:
Approx. 1 m/s

BATTERY:
Up to four
hours run time
from eight AAA
rechargeable
batteries

COMPATIBILITY:
Reversible
upper chassis
plate to
accommodate
Raspberry Pi 2,
3, 3B, 3B+, and
all Raspberry Pi
Zero models

PiBorg offers three-wheels of madcap mayhem
with its latest robot kit. By **Lucy Hattersley**

RockyBorg is a slightly madcap racing
robot from PiBorg, the geniuses behind
Formula Pi (Raspberry Pi's premier AI
racing competition). This latest creation follows
in the tyre tracks of MonsterBorg and DiddyBorg,
but with a significant reduction in price and
complexity, and one less wheel. Indeed. This is the
first three-wheeler we've tested at Raspberry Pi
towers. And we're entirely smitten.

Most of the robots we encounter have steerable
front wheels, or a tank-like differential steering
(where the speed or direction of the left and right
motors are adjusted to turn). RockyBorg does
things differently. Two 180 rpm motors on the rear
provide forward momentum, while a servo tilts
the whole body of the robot (including an affixed
front wheel) to change direction. The result is
a trike that tilts into corners. Everybody we've
shown RockyBorg to has loved it.

The on-board Camera Module (not supplied) tilts
with the body, leading to some great race footage.
This YouTube video shows RockyBorg in all its epic
action (**magpi.cc/wceWXb**).

RockyBorg is also good value at just £99.
This includes the acrylic parts, servo, and two
motors, plus the new custom RockyBorg motor
controller. You also get the on/off switch and
battery compartment (but need to supply your own
AAA batteries).

The supplied motor controller is similar to the
PiBorg's ZeroBorg, with support for the two direct
motors and a single servo.

You need to bring your own Raspberry Pi to the
party, plus a Camera Module if you want to add
vision to RockyBorg. It's not a complete kit, but
you can use it with most Raspberry Pi boards.

Ours was built with Raspberry Pi Zero WH, but
we've also tested one with Raspberry Pi 3A+. All
recent models of Raspberry Pi can be mounted to
the side, and PiBorg is developing an add-on for
Raspberry Pi 4 that adapts the ports on the newer
board and adds a 3 A power supply.

▶ Dual DC motors provide
forward propulsion, while
a servo motor tilts the
body to provide steering

> ❝ It's not a complete kit, but
> you can use it with most
> Raspberry Pi boards ❞

Building the RockyBorg

RockyBorg is a clever design. It features two
vertical plates around the front wheel, and a top
and bottom plate (both made of acrylic). These
clip together to box in the battery compartment,
keeping it secure and adding stability to the centre
of the robot.

Your Raspberry Pi board sits attached to the left plate, and the RockyBorg motor controller fits on top of Raspberry Pi. The wires from the motors feed through the back plate, through a hole in the top, and out another hole in the left.

It is a fiddly thing. Take a look at the RockyBorg build instructions in order to see what's in store (**magpi.cc/pTiCmO**). It's not a particularly difficult build, but be sure to follow the instructions carefully. There's a lot of looping of wires and cables that needs to be done with precision.

Once the robot is built, the wheels must be calibrated. PiBorg has a calibration guide (**magpi.cc/vGBtCQ**) which enables you to test each motor is working correctly and set the default return position for the robot.

Then it's on to the software installation (**magpi.cc/mOmMXC**), which pulls code from PiBorg's GitHub repo (**magpi.cc/hGtKhg**). The code enables remote control of RockyBorg with a Bluetooth controller (we tested it with a PS4 gamepad). And much fun is to be found jamming

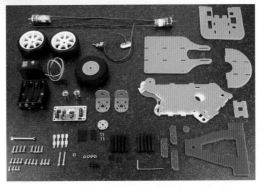

the RockyBorg around. RockyBorg also responds to a web interface, which provides camera feedback.

From there, you move on to the API interface. It's still a work in progress, and we've recently seen OpenCV support added, so hopefully you'll be able to use it to test out self-driving in the future. How easy this will be with the tilting camera will be interesting, though. Our experience with remote control and Python testing was fabulous. ⬚

◀ All the parts included with the RockyBorg build. You need to supply your own Raspberry Pi board and Camera Module

Verdict

We're letting some of the fiddly build complexity slide because we love playing with the end result so much. And, it's great value! RockyBorg is our favourite RC robot to date.

9/10

Enviro+

▶ Pimoroni ▶ **magpi.cc/ppkiiN** ▶ £45 / $48

Monitor your world with this all-in-one
environmental sensing board. By **Phil King**

SPECS

BUILT-IN SENSORS:
BME280 temperature/ pressure/ humidity, LTR559 light/ proximity, MEMS mic, MiCS-6814 gas

DISPLAY:
0.96-inch colour LCD (160×80)

OPTIONAL SENSOR:
Plantower PMS5003

Developed in conjunction with Dr Nate Adams, a molecular biologist at the University of Sheffield, the Enviro+ turns your Raspberry Pi into a complete environmental monitoring station. For this it features four built-in sensors, some of them multifunctional, so it can gather plenty of useful data, including for air quality. Not only that, but if using it in a headless Raspberry Pi setup, without a monitor, its tiny colour LCD screen offers a convenient way of displaying readings. There's also the option of plugging in a particular matter sensor (not included).

Like the earlier Enviro pHAT – reviewed back in *The MagPi* #49 (**magpi.cc/49**) and still available – the new board has a slimline pHAT form factor that matches Raspberry Pi Zero, although it can be used on any Raspberry Pi model. This time no soldering is required, as it comes with a female GPIO header attached.

Lacking the earlier board's motion sensors, the Enviro+ is intended purely for environmental monitoring. To this end, it incorporates a range of useful sensors.

Sensory overload

First up, a standard BME280 weather sensor is used to monitor temperature, barometric pressure, and humidity. This is positioned at the left edge of the board, away from the Raspberry Pi's CPU. Even so, you'll need to adjust its temperature reading (by measuring that of the CPU itself and deducting a factor of it).

A smartphone-style LTR-559 light and proximity sensor detects the ambient light level and also proves handy as a substitute for a push-button when you put your finger on it. A tiny MEMs microphone measures sound levels, useful for monitoring noise pollution, and can also be used to record audio.

Most notable is the inclusion of a MiCS6814 analogue gas sensor. This can detect three different groups of gases: reducing, oxidising, and NH_3 (ammonia). While levels of individual gases can't be discerned for the first two groups, the major ones are carbon monoxide (reducing) and nitrogen dioxide (oxidising).

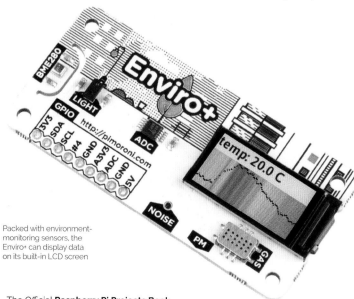

▶ Packed with environment-monitoring sensors, the Enviro+ can display data on its built-in LCD screen

▲ Available separately, the PMS5003 sensor measures the number of tiny particles – up to 1, 2.5, and 10 microns – in the air

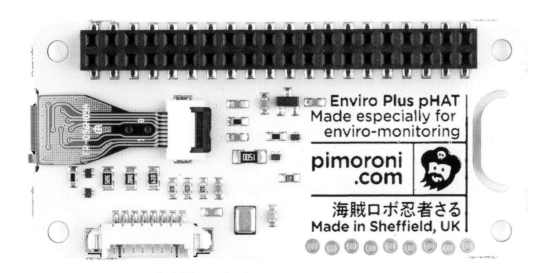

Near the gas sensor is a port to attach an optional particulate matter sensor, such as the Plantower PMS5003 (available separately for £25). This is used to measure numbers of tiny particles of sizes up to 1 micron (ultra-fine), 2.5 microns (combustion particles, organic compounds, metals), and 10 microns (dust, pollen, and mould spores). The board also features a nine-pin unpopulated header connected to selected GPIO pins. The finishing touch is the inclusion of a

Citizen science

Luftdaten is an open data project with a worldwide network of citizen scientists monitoring the air quality of their local environment – and, equipped with an Enviro+, you can become a part of it. Just run the **luftdaten.py** code example, register on the website (including your Raspberry Pi's displayed ID number), and you can start contributing your data – from the built-in BME680 weather sensor and add-on PMS5003 particulate matter sensor – which will then be shown on the site's world map.

0.96-inch colour LCD screen. It may be small, but it's ideal for displaying data out in the field, in a headless setup. It can even show some cool-looking scrolling graphs for live data, as shown in one of the Python code examples provided.

Environmental examples

Several code examples are included with the Enviro+ Python library for the board. Installation is simple enough, involving three terminal commands. The install script enables I²C, SPI, and serial interfaces on your Raspberry Pi, disables the serial console, and also enables a mini UART interface for the optional PMS5003 particulate matter sensor. If you ever need to revert this configuration change, there's an uninstall script.

The most impressive code example is **all-in-one.py**, which demonstrates most of the features of the board, taking readings from the various sensors (bar the mic) and displaying them in scrolling graph form on the mini LCD. To switch the latter between different readings, you simply tap the light sensor with your finger.

Another code example enables you to become a citizen scientist by uploading live data (from the BME280 and PMS5003) to the Luftdaten open-source air-quality monitoring project website (see 'Citizen science' box).

While the Enviro+ may seem a little pricey for a pHAT, it does cram a lot of useful sensors – which we reckon, if bought separately, would cost around £40 or more – into a handy package, along with that cool LCD screen to display your data. M

Verdict

If you want to create an air quality-monitoring project, this board is ideal, packing a raft of useful sensory tech into a small form factor, along with a handy LCD screen to display your data.

9/10

NanoSTEM IOT
Weather Kit

▶ Nanomesher ▶ **magpi.cc/rpPidc** ▶ £64 / $80

A much smaller weather kit solution, complete with Nanomesher's signature tiny display. **Rob Zwetsloot** braves the elements to test it

SPECS

DISPLAY:
1.3" High Contrast OLED (128×96 resolution)

PRESSURE SENSOR:
BMP180 (range 300–1100hPa)

HUMIDITY & TEMPERATURE SENSOR:
SHT31

AIR QUALITY SENSOR:
CCS811 (TVOC and eCO2 sensing)

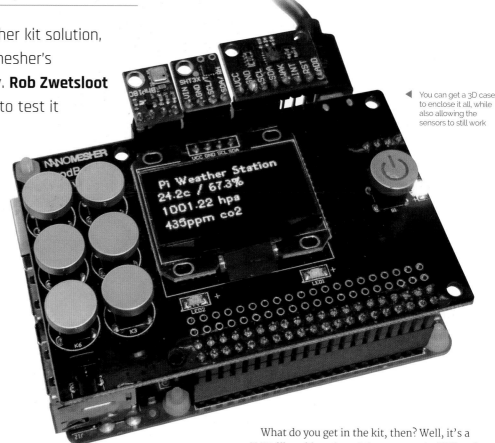

◀ You can get a 3D case to enclose it all, while also allowing the sensors to still work

Verdict

A great little IoT weather station that's easy to use and configure. It can be used around the house, or easily made portable with a mobile battery.

9/10

Home weather kits are excellent for a number of reasons. With the right calibration, they can be more accurate than a lot of basic solutions. In addition, by nature of being a computer add-on, they allow you to keep track of data over a long period of time – great for science, and also good for stuff like timing your heating in the winter, or even just kitchen conditions when cooking particularly fussy foods.

The NanoSTEM IoT Weather Kit is one of the smallest solutions we've seen for this, and it's relatively cheap as well. Featuring an at-a-glance display for basic data readouts and an online dashboard with the full suite of data, it's basically the full package.

What do you get in the kit, then? Well, it's a HAT-like add-on board that covers the USB and Ethernet ports, along with the rest of a Raspberry Pi. This is the ProdBoard that Nanomesher makes, which includes a tiny 1.3" OLED display, but adding a pressure sensor, humidity and temperature sensor, and an air quality sensor attached to the I²C ports.

Plug and play

There's also a microSD card that is all preconfigured – very welcome, as we know how tricky these sensors can be to set up. Air quality monitors aren't something we see regularly in Western weather kits, but it's a big thing in built-up areas of China. Everything is ready to go out of the box – even the sensors are plugged into the board – so as long as you have access to power, you can get very quickly started with monitoring. 🄼

Pi**Talk**

▶ SB Components ▶ **pitalk.co.uk** ▶ £58 / $76

▲ Being a standard HAT, installation takes minutes and the software configures Raspberry Pi for you

Turn your Raspberry Pi into a 4G mobile phone with the PiTalk HAT from SB Components. **PJ Evans** keeps talking

Over the past few years, we've been lucky to see Raspberry Pi get more and more communication options. It's no surprise then that there's been great interest in getting Raspberry Pi devices onto the mobile data network. SB Components' PiTalk range does just that, except rather than stopping at just a data-capable device, the PiTalk HAT turns your Raspberry Pi into a fully-fledged smartphone with voice and SMS support.

Just add SIM

The HAT features a Quetec SoC that adds everything Raspberry Pi needs to get on the mobile data network; you only need to supply a micro SIM. Our tests with EE met with failure, but a Vodafone SIM worked first time. As the HAT only uses serial communications, nearly all the GPIO pins remain unused and the HAT has 'through' pins for further

expansion. SB offers a range of small touchscreens that can be added for a more phone-like experience.

To get everything running, software is supplied, but this is squarely aimed at the touchscreen. Although we were able to make voice calls, send SMS messages, and transmit data, the interface is poor to unusable without a touchscreen.

It is not an end-user product, but rather something on which other projects can be based.

🞂 Perfect if you are interested in remote automation 🞀

We were able to find an example project using Python to exchange SMS messages and trigger GPIO pins in response. It's great for experimenting with smartphone technology, and perfect if you are interested in remote automation. 🅼

SPECS

MOBILE COMMUN- ICATIONS:
UMTS/HSDPA and GSM/ GPRS/EDGE

SPEAKER / MIC:
On-board jumpers and 3.5-inch connector

LOCATION SERVICES:
Quec Locator. Positioning based on cell tower

▼ The smartphone interface could be improved for desktop monitors, but did work as advertised

Verdict

While let down by a tricky interface, the PiTalk is ideal for makers planning remote monitoring projects. It's better combined with a touchscreen.

7/10

10 Best:
Raspberry Pi HATs

The very best Hardware Attached on Top for your Raspberry Pi

HATs are incredible add-ons to a Raspberry Pi that increase its functionality in a huge number of ways – from added sensors and inputs for fun projects, to practical applications in business and enterprise. Here are some of the best… 𝔐

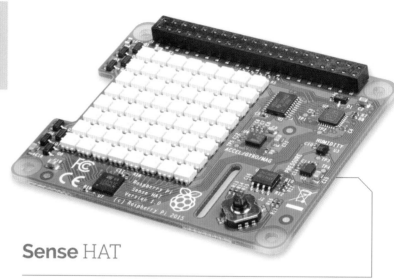

Sense HAT

Space-faring sensor

Used in the Astro Pi devices up on the International Space Station, this cool HAT has an 8×8 pixel display, environmental sensors, accelerometer, and a little joystick. There's loads of great Raspberry Pi resources that use it as well.

▶ £32 / $40
▶ **magpi.cc/BsVbhG**

GFX HAT

Display and inputs

We really like the GFX HAT – not only is it a very nice display that is easy to program for, it also has a series of (capacitive touch) buttons. It's amazing for practical projects or any Raspberry Pi that's turned on a lot.

▶ £22 / $23
▶ **magpi.cc/ZWvcLG**

PoE HAT

Power over Ethernet

Raspberry Pi is popular in enterprise settings, and the PoE HAT offers a more efficient way to add Raspberry Pi boards to a system. With PXE boot and power from the Ethernet port, you can do a lot with a 3B+.

▶ £18 / $20
▶ **magpi.cc/aqpwZc**

Piano HAT

Tickle the ivories

A lot of HATs add important and useful functions, but others are just for fun, like the surprisingly fully featured Piano HAT.

▶ £16 / $17
▶ **magpi.cc/iqPELX**

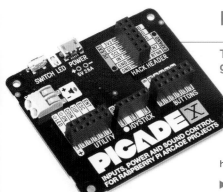

Picade X HAT

Totally awesome video games

Inside the Picade is this amazing control HAT that's purpose-built for Raspberry Pi arcade machines. It includes audio and inputs, along with other bits and pieces, and it's the perfect heart for your arcade build.

▶ £16 / $17
▶ **magpi.cc/BupAFF**

PaPiRus HAT

E-ink display

You can make anything look 20% classier with an e-ink display. Trust us, we've got the science to back it up. It's very low-power, depending on usage, and looks great even in bright sunlight.

▶ £41 / $54
▶ **magpi.cc/ikcQsi**

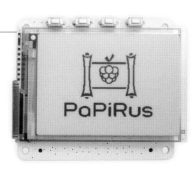

HiFiBerry DAC+ DSP

Powerful audio amplifier

This powerful DAC also includes a DSP (digital signal processor) which allows for full control of the way your audio is output from Raspberry Pi. HiFiBerry calls it a 'DAC on steroids'.

▶ £60 / $78
▶ **magpi.cc/EsHtjE**

Flick HAT

Gesture control

The Flick HAT can handle all of your gesture control needs, with 3D tracking that lets you control Raspberry Pi with some simple gestures, even from a moderate distance (about 15 cm away).

▶ £20 / $26
▶ **magpi.cc/vXZLzF**

Unicorn HAT HD

High-def LEDs

The wonderful Unicorn HAT HD lets you create amazing multicoloured visuals, perfect for a disco or project that needs a dot-matrix display aesthetic. It fits in a lot of cases as well.

▶ £34 / $37
▶ **magpi.cc/uitfMn**

TV HAT

Must-see TV

Tune into digital TV in Europe with this official HAT that lets you stream the signal around the home. It's the perfect complement to a Kodi box, and it sits nicely on top of a Raspberry Pi Zero.

▶ £20
▶ **magpi.cc/ryviXi**

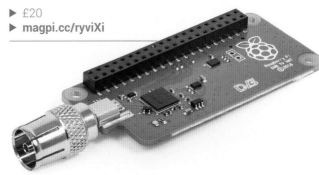

HAT DEFINITION

Introduced in 2014 when Raspberry Pi B+ was released, the HAT standard relies on the use of two dedicated GPIO pins on a 40-pin Raspberry Pi that allow for automatic configuration from the HAT so Raspberry Pi can use it. Learn more specifics here: **magpi.cc/PAHaGk**

10 Best: Home automation add-ons

Get Raspberry Pi controlling your home with these IoT gadgets

The dream of home automation is forever present among makers and other tech types, and Raspberry Pi has helped many people turn their Enterprise fantasies into a reality. Here are some kits, add-ons, and other gadgets that can help. ▨

Automation HAT

All-in-one automation

If you have big home automation plans, or already have some serious IoT around your house, you'll want to look into the Automation HAT: you can plug a lot of devices into it.

▶ £29 / $31
▶ **magpi.cc/fgMGmr**

Energenie Pi-mote

Remote-control plug

A quick and easy way to do some home automation is to remotely control a mains socket with your Raspberry Pi and some code! The Energenie Pi-mote allows you to do just that.

▶ £17 / $18
▶ **magpi.cc/FepLDV**

SparkFun ESP32 Thing

WiFi smart home

ESP32 is a standard that lets you use wireless LAN to communicate with the various IoT/home automation projects you've set up. This one also has Bluetooth as part of it!

▶ £21 / $23
▶ **magpi.cc/UchWDj**

Google AIY Voice Kit

Vocal commands

We had this as a freebie with the magazine once – the AIY Voice Kit allows you to add powerful voice control to your Raspberry Pi, and any connected IoT or home automation systems in the process.

▶ £25 / $20
▶ **magpi.cc/xDJDPp**

Pi NoIR Camera V2

See in the dark

If you're setting up a CCTV network in your home, or want a front-door camera that works 24 hours a day, then the IR version of the Raspberry Pi Camera Module is for you.

▶ £25 / $25
▶ **magpi.cc/ircamera**

Gravity
Light Sensor

Ambient light control

Ambient light sensors are very common (you probably have one in your phone) and can be a good way to slowly bring up lights, as it gets dark outside, in a more natural way.

▶ £7 / $9
▶ **magpi.cc/ibSTjk**

Amazon
Dash Button

Press for anything

A favourite among hackers is the Amazon Dash Button – you can easily use it as a remotely connected button that does what you'd like it to do. And pretty cheaply as well!

▶ £5 / $5
▶ **magpi.cc/DE2xBE**

SparkFun
OpenPIR

Motion sensing

Want to trigger a camera recording? Or lights turning on? Or literally anything involving seeing if something has moved in a location? A PIR sensor like this will help.

▶ £15 / $16
▶ **magpi.cc/rcdvXL**

AUTOMATION SOFTWARE

While you can do a lot using Raspbian, there are dedicated home automation operating systems for Raspberry Pi that are already preconfigured. We like openHAB, which you can find here: **openhab.org**.

DHT22
temperature-humidity sensor

Thermostats beware

With stuff like Nest around, it shouldn't be too surprising how many people use a Raspberry Pi and a temperature sensor such as this as a thermostat for their house. As an Astro Pi project demonstrated, you can use the humidity and temperature parts together to detect a person.

▶ £10 / $13
▶ **magpi.cc/PZrpfv**

Philips Hue Lights

Controllable bulbs

An excellent solution for controlling your lights is to have remote-controllable light bulbs like the Hue! Find our tutorial on how to control them with a Raspberry Pi in *The MagPi* #61 (**magpi.cc/61**).

▶ Various
▶ **meethue.com**

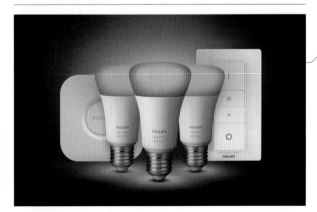

10 Best:

Laptop kits & projects

Want to make your Raspberry Pi more portable? Here are some of the best ways...

Raspberry Pi devices powering old laptops and custom mini computers have always been a popular idea in the Raspberry Pi community. As tech compatible with Raspberry Pi and other hardware becomes more available, so has the feasibility of this kind of project – so much so, you can now get all-in-one computer kits. Here are some of our faves. 𝕄

pi-top 3

True original

This lean, green, Kickstarted module laptop machine is the original Raspberry Pi laptop kit. There are several variations of it that sell for a wide variety of prices, but the original laptop configuration remains a favourite to this day.

▶ £245 / $320
▶ **pi-top.com**

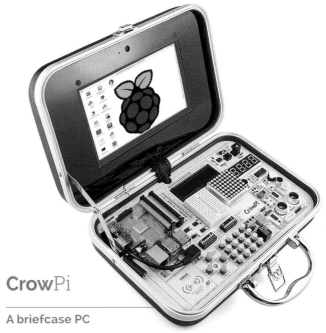

CrowPi

A briefcase PC

Another crowdfunded laptop, the CrowPi looks like a spy kit worthy of Bond himself. Full of cool components and connectivity, it's a great way to carry fun Raspberry Pi-powered projects around.

▶ £186 / $239
▶ **magpi.cc/DcAWER**

Computer Kit Touch

Screen and keyboard, no folding

Kano has been producing Raspberry Pi-powered computer kits for years. This one is a touchscreen tablet with an additional external keyboard. While not a traditional style of laptop, it functions similarly to one.

▶ £280 / $280
▶ **magpi.cc/gLpTVF**

RasPad

Raspberry Pi tablet

This tablet computer kit is an interesting implementation of Raspberry Pi: one part tablet, one part mobile-powered Raspberry Pi. You can easily hook it up to an HDMI display and any of your usual USB inputs at any time.

▶ £238 / $259
▶ **magpi.cc/wpZiSS**

LapPi

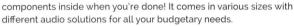

See-through DIY

This recent project from the folks at SB Components is a laptop that you not only build, but also see all the components inside when you're done! It comes in various sizes with different audio solutions for all your budgetary needs.

▶ From £119 / $155
▶ **magpi.cc/igqsAs**

Mini Handheld Notebook

Build from scratch

This Adafruit project from the venerable Ruiz Bros lets you create a very tiny Raspberry Pi laptop/netbook using a full-sized Raspberry Pi. You'll need to source the parts individually, although that does mean you can switch them for alternatives as you see fit.

▶ **magpi.cc/vnCQJR**

Portable Laptop

Barebones build

Designed by Pi Supply to make use of its versatile PiJuice battery module, this very simple build uses the bare minimum of elements to make a laptop-like design possible. And it works just as well as the Kano and RasPad kits do!

▶ **magpi.cc/SzyDbg**

Nano Pi2 UMPC

Palm-sized laptop

This is the smallest 'laptop' we've seen that makes use of a full-sized Raspberry Pi. It even features a neat clamshell design! Again, you'll have to source all the parts to make it yourself. It's well worth the effort, though.

▶ **magpi.cc/TUULny**

Piper Computer Kit 2

The play-chest computer

Another educational computer kit, the Piper computer is laid out like a Raspberry Pi-powered laptop but uses the space more efficiently for storage and setting up fun programming tasks. You can just install Raspbian on it, though, if you wish.

▶ £274 / $299
▶ **playpiper.com**

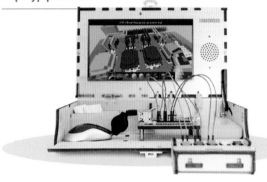

Lego Raspberry PiBook

Dig out your bricks

An oldie but a goodie – while this project was designed with the original Raspberry Pi Model B in mind (with an updated Raspberry Pi Zero version later), the setup is very easy to modify so that you can use a more powerful Raspberry Pi 3B+ or 4.

▶ **magpi.cc/FJawXH**

LAPTOP BUILDS AND IDEAS!

If you fancy trying out some other laptop building methods, take a look at *The MagPi* #74 (**magpi.cc/74**) for our 'Build a laptop' feature!